The MEAP Solution:

A Technological Approach

Written by
Twianie Mathis, Ed.D

Mathematical Solutions Publishing Company
P.O. Box 36365
Grosse Pointe Farms, Michigan 48236-0365

ISBN 0-9718019-0-8: $25.00

Acknowledgments

Special Thanks to God, with whom, all things are possible, my husband Cornell, Michelle, Bettie, Anita, Jean and Vivian for their positive support and encouragement.

To Dr. Calloway, Mrs. Estell, Mr. Kelly, Ms. Jackson, Mr. Sada, Mrs. Smith, Mr. Campbell and Dr. Theodore Roosevelt Sykes, the mathematics teachers who worked tirelessly to develop the concepts and constructs within me.

Credits /Disclaimers

TI-83™ and TI-83 Plus™ likeness and sequences reprinted from the TI-83 and TI-83 Plus Guidebooks. This was done with permission from Texas Instruments, Dallas, Texas.

Cover designed by Victor Green of "Complete Graphics", Detroit, Michigan, 1-313- 475-1692.

Model Assessment Problems, Reference Sheet and MICLiMB information printed with permission from the Michigan Department of Education.

Clipart derived from Windows 98™

Dictionary Terms taken from "Webster's New World Dictionary, Third College Edition" and "American Standard Dictionary".

Names, characters, places, and incidents are either products of the author's imagination or are used fictiously. Any resemblance to actual events, locales, or persons, living or dead, is entirely coincidental.

Mathematical Solutions Publishing Company makes no warranty, either expressed or implied, including but not limited to any implied warranties of merchantability and fitness for a particular purpose, regarding any programs or book materials and makes such materials available solely on an "as-is" basis.

In no event shall Mathematical Solutions Publishing Company be liable to anyone for special, collateral, incidental, or consequential damages in connection with or arising out of the purchase or use of these materials, and the sole and exclusive liability of Mathematical Solutions Publishing Company, regardless of the form of action, shall not exceed the purchase price of this book. Moreover, Mathematical Solutions Publishing Company shall not be liable for any claim whatsoever against the use of these materials by any other party.

Table of Contents

Introduction

"The MEAP SOLUTION: A Technological Approach", serves as a technologically based resource tool for the mathematics portion of the Michigan Educational Assessment Program (MEAP) at the high school level.

Each section consists of lessons that correspond to the strands, standards and benchmarks as noted within the Michigan Curriculum Frameworks. Lessons are addressed by strand. The lessons begin with terms associated with each strand and are presented in two formats.

1. Multiple choice questions
2. Constructed response questions

© 2002

Mathematical Solutions Publishing Company
P.O. Box 36365
Grosse Pointe Farms, Michigan 48236-0365

The MEAP Solution Utilizes
The TI-83™ or ...

TI-83 Plus™

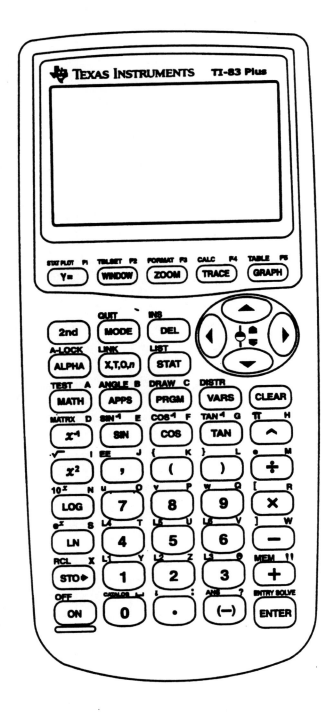

Michigan Curriculum Framework 9-12 Mathematics Strands, Standards, and Benchmarks

<u>Strands</u>

The content has been categorized into the following six strands:

1. **Patterns, Relationships and Functions**
   Students recognize similarities among objects and events, generalize patterns and relationships, and use them to describe the physical world, to explain variation, and to make predictions and solve problems.

2. **Geometry and Measurement**
   Students use analytical and spatial concepts of shape, size, position, measurement and dimension to understand and interpret the three-dimensional world in which we live.

3. **Data Analysis and Statistics**
   Students organize, interpret and transform data into useful knowledge to make predictions and decisions based on data.

4. **Number Sense and Numeration**
   Students quantify and measure objects, estimate mathematical quantities, and represent and communicate ideas in the language of mathematics.

5. **Numerical and Algebraic Operations and Analytical Thinking**
   Students represent quantitative situations with numerical and algebraic symbolism and use analytic thinking to solve problems in significant contexts and applications.

6. **Probability and Discrete Mathematics**
   Students deal with uncertainty, make informed decisions based on evidence and expectations, exercise critical judgment about conclusions drawn from data, and apply mathematical models to real-world phenomena.

Wording taken directly from the "Michigan Curriculum Frameworks"

Mathematics Standards

The following 15 mathematics standards have been written within the six strands as follows:

Patterns, Relationships and Functions

1. Patterns – Students recognize similarities and generalize patterns to create models and make predictions, describe the nature of patterns and relationships, and construct representations of mathematical relationships.

2. Variability and Change – Students describe the relationships among variables, predict what will happen to one variable as another variable is changed, analyze natural variation and sources of variability, and compare patterns of change.

Geometry and Measurement.

3. Shape and Shape Relationships – Students develop spatial sense, use shape as an analytic and descriptive tool, identify characteristics and define shape, identify properties and describe relationships among shapes.

4. Position – Students identify locations of objects, identify location relative to other objects, and describe the effects of transformations (e.g., sliding, flipping, turning, enlarging, reducing) on an object.

5. Measurement – Students compare attributes of two objects or of one object with a standard (unit), and analyze situations to determine what measurement(s) should be made and to what level of precision.

Data Analysis and Statistics

6. Collection, Organization and Presentation of Data – Students collect and explore data, organize data into a useful form, and develop skill in representing and reading data displayed in different formats.

7. Description and Interpretation – Students examine data and describe characteristics of a distribution, relate data to the situation in which they arose, and use data to answer questions convincingly and persuasively.

8. Inference and Prediction – Students draw defensible inferences about unknown outcomes, make predictions, and identify the degree of confidence they have in their predictions.

Wording taken directly from the "Michigan Curriculum Frameworks"

Number Sense and Numeration

9. Concepts and Properties of Numbers – Students experience counting and measuring activities to develop intuitive sense about numbers, develop understanding about properties of numbers, understand the need for and existence of different sets of numbers, and investigate properties of special numbers.

10. Representation and Uses of Numbers - Students recognize that numbers are used in different ways such as counting, measuring, ordering and estimating, understand and produce multiple representations of a number, and translate among equivalent representations.

11. Number Relationships – Students investigate relationships such as equality, inequality, inverses, factors and multiples, and represent and compare very large and very small numbers

Numerical and Algebraic Operations and Analytic Thinking

12. Operations and their Properties – Students understand and use various types of operations (e.g., addition, subtraction, multiplication, division) to solve problems.

13. Algebraic and Analytic Thinking – Students analyze problems to determine an appropriate process for solution, and use algebraic notations to model or represent problems.

Probability and Discrete Mathematics

14. Probability – Students develop an understanding of the notion of certainty and of probability as a measure of degree of likelihood that can be assigned to a given event based on the knowledge available, and make critical judgments about claims that are made in probabilistic situations.

15. Discrete Mathematics – Students investigate practical situations such as scheduling, routing, sequencing, networking, organizing and classifying, and analyze ideas like recurrence relations, induction, iteration, and algorithm design

Wording taken directly from the "Michigan Curriculum Frameworks"

8

Benchmarks "•"

Patterns, Relationships and Functions

1. Patterns – Students recognize similarities and generalize patterns to create models and make predictions, describe the nature of patterns and relationships, and construct representations of mathematical relationships.

- Analyze and generalize mathematical patterns including sequences, series and recursive patterns.
- Analyze, interpret and translate among representations of patterns including tables, charts, graphs, matrices and vectors.
- Study and employ mathematical models of patterns to make inferences, predictions and decisions.
- Explore patterns (graphic, numeric, etc.) characteristics of families of functions; explore structural patterns within systems of objects, operations or relations.
- Use patterns and reasoning to solve problems and explore new content.

2. Variability and Change – Students describe the relationships among variables, predict what will happen to one variable as another variable is changed, analyze natural variation and sources of variability, and compare patterns of change.

- Identify and describe the nature of change and begin to use the more formal language such as rate of change, continuity, limit, distribution and deviation.
- Develop a mathematical concept of function and recognize that functions display characteristic patterns of change (e.g., linear, quadratic, exponential)
- Expand their understanding of function to include non-linear functions, inverses of functions, and piece wise and recursively-defined functions.
- Represent functions using symbolism such as matrices, vectors, and functional representation (f(x)).
- Differentiate and analyze classes of functions including linear, power, quadratic, exponential, circular and trigonometric functions and realize that many different situations can be modeled by a particular type of function.
- Increase their use of functions and mathematical models to solve problems in context.

Geometry and Measurement

3. Shape and Shape Relationships – Students develop spatial sense, use shape as an analytic and descriptive tool, identify characteristics and define shape, identify properties and describe relationships among shapes.

- Use shape to identify plane and solid figures, graphs, loci, functions, and data distributions.

Wording taken directly from the "Michigan Curriculum Frameworks"

- Determine necessary and sufficient conditions for the existence of a particular shape and apply those conditions to analyze shapes.
- Use transformational, coordinate or synthetic methods to verify (prove) the generalizations they have made about properties of classes of shapes.
- Draw and construct shapes in two and three dimensions and analyze and justify the steps of their constructions.
- Study transformations of shapes using isometries, size transformations and coordinate mappings.
- Compare and analyze shapes and formally establish the relationships among them, including congruence, similarity, parallelism, perpendicularity and incidence.

4.Position – Students identify locations of objects, identify location relative to other objects, and describe the effects of transformations (e.g., sliding, flipping, turning, enlarging, reducing) on an object.
- Locate and describe objects in terms of their position, including polar coordinates, three dimension Cartesian coordinates, vectors, and limits.
- Locate and describe objects in terms of their orientation and relative position, including displacement (vectors), phase shift, maxima, minima and inflection points; give precise mathematical descriptions of symmetries.
- Give precise mathematical descriptions of transformations and describe the effects of transformations on size, shape, position and orientation.
- Describe the locus of a point by a rule or mathematical expression; trace the locus of a moving point.
- Use concepts of position, direction and orientation to describe the physical world and to solve problems.

5.Measurement – Students compare attributes of two objects or of one object with a standard (unit), and analyze situations to determine what measurement(s) should be made and to what level of precision.
- Select and use appropriate tools; make accurate measurements using both metric and common units, and measure angles in degrees and radians.
- Continue to make and apply measurements of length, mass (weight), time, temperature, area, volume, angle; classify objects according to their dimensions.
- Estimate measures within a specified degree of accuracy and evaluate measurements for accuracy, precision and tolerance.
- Interpret measurements and explain how changes in one measurement may affect other measures.
- Use proportional reasoning and indirect measurements, including applications of trigonometric ratios, to measure inaccessible distances and to determine derived measures such as density.

Wording taken directly from the "Michigan Curriculum Frameworks"

- Apply measurement to describe the real world and to solve problems.

Data Analysis and Statistics

6.Collection, Organization and Presentation of Data – Students collect and explore data, organize data into a useful form, and develop skill in representing and reading data displayed in different formats.
- Collect and explore data through observation, measurement, surveys, sampling techniques and simulations.
- Organize data using tables, charts, graphs, spreadsheets and data bases.
- Present data using the most appropriate representation and give a rationale for their choice; show how certain representations may skew the data or bias the presentation.
- Identify what data are needed to answer a particular question or solve a given problem, and design and implement strategies to obtain, organize and present those data.

7.Description and Interpretation – Students examine data and describe characteristics of a distribution, relate data to the situation in which they arose, and use data to answer questions convincingly and persuasively.
- Critically read date from tables, charts or graphs and explain the source of the data and what the data represent.
- Describe the shape of a data distribution and determine the measures of central tendency, variability and correlation.
- Use the data and their characteristics to draw and support conclusions.
- Critically question the sources of data; the techniques used to collect, organize and present data; the inferences drawn from the data; and the sources of bias and measures taken to eliminate such bias.
- Formulate questions and problems, and interpret data to answer those questions.

8.Inference and Prediction – Students draw defensible inferences about unknown outcomes, make predictions, and identify the degree of confidence they have in their predictions.
- Make and test hypothesis.
- Design investigations to model and solve problems; also employ confidence intervals and curve fitting in analyzing the data.
- Formulate and communicate arguments and conclusions based on data and evaluate their arguments and those of others.
- Make predictions and decisions based on data, including interpolations and extrapolations.
- Employ investigations, mathematical models and simulations to make inferences and predictions to answer questions and solve problems.

Wording taken directly from the "Michigan Curriculum Frameworks"

Number Sense and Numeration

9. Concepts and Properties of Numbers – Students experience counting and measuring activities to develop intuitive sense about numbers, develop understanding about properties of numbers, understand the need for and existence of different sets of numbers, and investigate properties of special numbers.
- Develop an understanding of irrational, real and complex numbers.
- Use the (a+bi) and polar forms of complex numbers.
- Develop an understanding of the properties of the real and complex number systems and of the properties of special numbers including p, i, e, and conjugates.
- Apply their understanding of number systems to model and solve mathematical and applied problems.

10. Representation and Uses of Numbers - Students recognize that numbers are used in different ways such as counting, measuring, ordering and estimating, understand and produce multiple representations of a number, and translate among equivalent representations.
- Give decimal representations of rational and irrational numbers and coordinate and vector representations of complex numbers.
- Develop an understanding of more complex representations of numbers, including exponential and logarithmic expressions, and select an appropriate representation to facilitate problem solving.
- Determine when to use rational approximations and the exact values of numbers such as e, p and the irrationals.
- Apply estimation in increasingly complex situations.
- Select appropriate representations for numbers, including representation of rational and irrational numbers and coordinate or vector representation of complex numbers, in order to simplify and solve problems.

11. Number Relationships – Students investigate relationships such as equality, inequality, inverses, factors and multiples, and represent and compare very large and very small numbers.
- Compare and order real numbers and compare rational approximations to exact values.
- Express numerical comparisons as ratios and rates.
- Extend the relationships of primes, factors, multiples and divisibility in an algebraic setting.

Wording taken directly from the "Michigan Curriculum Frameworks"

- Express number relationships using positive and negative rational exponents, logarithms, and radicals.
- Apply their understanding of number relationships in solving problems.

Numerical and Algebraic Operations and Analytic Thinking

12. Operations and their Properties – Students understand and use various types of operations (e.g., addition, subtraction, multiplication, division) to solve problems.
   - Present and explain geometric and symbolic models for operations with real and complex numbers, and algebraic expressions.
   - Compute with real numbers, complex numbers, algebraic expressions, matrices and vectors using technology and, for simple instances, with paper-and-pencil algorithms.
   - Describe the properties of operations with numbers, algebraic expressions, vectors and matrices and make generalizations about the properties of given mathematical systems.
   - Efficiently and accurately apply operations with real numbers, complex numbers, algebraic expressions, matrices, and vectors in solving problems.

13. Algebraic and Analytic Thinking – Students analyze problems to determine an appropriate process for solution, and use algebraic notations to model or represent problems.
   - Identify important variables in a context, symbolize them, and express their relationships algebraically.
   - Represent algebraic concepts and relationships with matrices, spreadsheets, diagrams, graphs, tables, physical models, vectors, equations and inequalities; and translate among the various representations.
   - Solve linear equations and inequalities algebraically and non-linear equations using graphing, symbol-manipulating or spreadsheet technology; and solve linear and non- linear systems using appropriate methods.
   - Analyze problems that can be modeled by functions, determine strategies for solving the problems, and evaluate the adequacy of the solutions in the context of the problems.
   - Explore problems that reflect the contemporary uses of mathematics in significant contexts and use the power of technology and algebraic and analytic reasoning to experience the ways mathematics is used in society.

Probability and Discrete Mathematics

14. Probability – Students develop an understanding of the notion of certainty and of probability as a measure of degree of likelihood that can be assigned to a given event based on the knowledge available, and make critical judgments about claims that are made in probabilistic situations.
   - Develop an understanding of randomness and chance variation and describe chance and certainty in the language of probability.

Wording taken directly from the "Michigan Curriculum Frameworks"

13

- Give a mathematical definition of probability and determine the probabilities of more complex events, and generate and interpret probability distributions.
- Analyze events to determine their dependence or independence and calculate probabilities of compound events.
- Use sampling and simulations to determine empirical probabilities and, when appropriate, compare them to the corresponding theoretical probabilities; understand and apply the law of large numbers.
- Conduct probability experiments and simulations to model and solve problems, including compound events.

15. Discrete Mathematics – Students investigate practical situations such as scheduling, routing, sequencing, networking, organizing and classifying, and analyze ideas like recurrence relations, induction, iteration, and algorithm design.
- Derive and use formulas for calculating permutations and combinations.
- Use sets and set relationships to represent algebraic and geometric concepts.
- Use vertex-edge graphs to solve network problems such as finding circuits, critical paths, minimum spanning trees, and adjacency matrices.
- Analyze and use discrete ideas such as induction, iteration and recurrence relations.
- Describe and analyze efficient algorithms to accomplish a task or solve a problem in a variety of contexts including practical, mathematical and computer-related situations.
- Use discrete mathematics concepts as described above to model situations and solve problems; and look for whether or not there is a solution (existence problems), determine how many solutions there are (counting problems), and decide upon a best solution (optimization problems).

Wording taken directly from the "Michigan Curriculum Frameworks"

MEAP 2002
Benchmarks: Restrictions and Recommendations
High School Mathematics

I.   Patterns, Relationships and Functions

Content Standard 1: (Patterns)
Students recognize similarities and generalize patterns, use patterns to create models and make predictions, describe the nature of patterns and relationships, and construct representations of mathematical relationships.

High School Benchmark 1: Analyze and generalize mathematical patterns including sequences, series, and recursive patterns.
MEAP Restrictions: Keep series to finite sums and restrict series to arithmetic and geometric. Students should engage in activities in which they use sequences, recursive patterns, algebraic notation and technology to show recognition of, generalization, of and descriptions of patterns.

High School Benchmark 2: Analyze, interpret and translate among representations of patterns including tables, charts, graphs, matrices and vectors.
MEAP Restrictions: Keep series to finite sums and restrict series to arithmetic and geometric. Exclude vectors from testing. Matrices will be assessed in context, and will be restricted to addition, subtraction, and scalar multiplication. Use of a graphing calculator is appropriate. Students should engage in activities in which they translate between various representations of mathematical patterns and real world phenomena.

High School Benchmark 3: Study and employ mathematical models of patterns to make inferences, predictions and decisions.
MEAP Restriction: Keep series to finite sums and restrict series to arithmetic and geometric. Students should engage in activities in which they make predictions and decisions based on mathematical models such as tables, graphs, charts, and equations.

High School Benchmark 4: Explore patterns (graphic, numeric, etc.) characteristic of families of functions; explore structural patterns within systems of objects, operations or relations.
MEAP Restriction: Keep series to finite sums and restrict series to arithmetic and geometric. Students should engage in which activities which they recognize, explore and classify families of functions and operations among functions.

High School Benchmark 5: Use patterns and reasoning to solve problems and explore new content.

MEAP Restriction: Keep series to finite sums and restrict series to arithmetic and geometric. Students should engage in activities in which they use patterns and reasoning to recognize and characterize real world phenomena. Students should explore new content to make inferences, predictions and decisions. Patterns will be presented as series, sequences, recursive patterns, tables, charts and graphs.

Content Standard 2: (Variability and Change)

Students describe the relationships among variables, predict what will happen to one variable as another variable is changed, analyze natural variation and sources of variability, and compare patterns of change.

High School Benchmark 1: Identify and describe the nature of change and begin to use the more formal language such as rate of change, continuity, limit, distribution and deviation.

MEAP Restriction: Exclude continuity and limits from testing. Restrict deviations to deviations from the mean. Graphing calculator appropriate. Students should engage in activities in which they identify and describe function behavior. Activities should include rate of change (slope), continuity and asymmetric behavior. Statistical function characteristics will include distribution and deviation from the mean and quartiles.

High School Benchmark 2: Develop a mathematical concept of function and recognize that functions display characteristic patterns of change (e.g., linear, quadratic, exponential).

MEAP-Like Item: Students should engage in activities in which they understand that families of functions have similar characteristics. Students should be able to identify and explain changes within a family, and describe differences among families.

High School Benchmark 3: Expand their understanding of function to include non-linear functions, composition of functions, inverses of functions, and piecewise- and recursively- defined functions.

MEAP Restriction: THIS BENCHMARK IS NOT CURRENTLY ASSESSED. Students may be expected to demonstrate an understanding of piecewise, recursively defined and inverse functions. Compositions of functions will be tested only at introductory levels.

High School Benchmark 4: Represent functions using symbolism such as matrices, vectors and functional representation ($f(x)$).

MEAP Restrictions: Exclude vectors from testing. The notation "f(x)" will be used in context only. Matrix operations of addition, subtraction, and scalar multiplication will be tested. Students should engage in activities in which they read, use and understand *f(x)* notation. Students may be expected to understand and use matrix and vector representation of functions.

High School Benchmark 5: Differentiate and analyze classes of functions including linear, power, quadratic, exponential, circular, and trigonometric functions, and realize that many different situations can be modeled by a particular type of function.
MEAP Restrictions: Keep exponents and powers simple. Limit trigonometric functions to sine, cosine, and tangent. Use right triangle definitions of trigonometric functions only. Graphing calculator appropriate. Students should engage in activities in which they use and understand characteristics of families of functions and their application in describing statistical data. Exponential and power functions will be tested only at an introductory level. Trigonometric functions will be restricted to sine, cosine and tangent for the current test.

High School Benchmark 6: Increase their use of functions and mathematical models to solve problems in context.
MEAP Restriction: Exclude continuity and limit. Restrict deviations to deviations from the mean. Exclude vectors from testing. *f(x)* notation will be used in context only. Keep exponents and powers simple. Limit trigonometric functions to sine, cosine, and tangent. Use right triangle definitions of trigonometric functions only. Graphing calculator appropriate. Students should engage in activities in which they use all of the functions previously described in problem solving situations. The same restrictions listed previously will be followed.

II.   Geometry and Measurement

<u>Content Standard 1: (Shape and Shape Relationships)</u>
Students develop spatial sense, use shape as an analytic and descriptive tool, identify characteristics and define shapes, identify properties and describe relationships among shapes.

High School Benchmark 1: Use shape to identify plane and solid figures, graphs, loci, functions and data distributions.
MEAP-Like Item: Students should engage in activities in which they may be asked about common names and properties of 2- and 3-dimensional shapes; linear, quadratic and exponential functions and their graphs; and data displays including bar graphs, histograms, and scatterplots.

High School Benchmark 2: Determine necessary and sufficient conditions for the existence of a particular shape and apply those conditions to analyze shapes.
MEAP-Like Item : Students should engage in activities in which they determine necessary and sufficient conditions for existence of a shape and analyze shape using those conditions. Students may be asked to name properties which guarantee a shape or properties characteristic of a given shape, and to draw conclusions.

High School Benchmark 3: Use transformational, coordinate or synthetic methods to verify (prove) the generalizations they have made about properties of classes of shapes.

MEAP-Like Item: Students should engage in activities in which they use transformational, coordinate, and synthetic methods to form conclusions.

High School Benchmark 4: Draw and construct shapes in two and three dimensions and analyze and justify the steps of their constructions.
MEAP-Like Item: The students should engage in activities in which they analyze the drawings of 2-dimensional shapes and 3-dimensional shapes and nets.

High School Benchmark 5: Study transformations of shapes using isometries, size transformations and coordinate mappings.
MEAP-Like Item: Students should engage in activities in which they calculate results of a transformation on known points or shapes and identify a transformation that produces a given result.

High School Benchmark 6: Compare and analyze shapes and formally establish the relationships among them, including congruence, similarity, parallelism, perpendicularity, and incidence.
MEAP-Like Item: Students should engage in activities in which they compare and analyze 2- and 3-dimensional shapes using congruence and similarity, parallels and perpendiculars, and intersections.

High School Benchmark 7: Use shape, shape properties and shape relationships to describe the physical world and to solve problems.
MEAP-Like Item: Students should engage in activities in which they use shapes, shape properties, and shape relationships to solve problems usually set in real world context.

Content Standard 2: (Position)
Students identify locations of objects, identify location relative to other objects, and describe the effects of transformations e.g., sliding, flipping, turning, enlarging, reducing on an object.

High School Benchmark 1: Locate and describe objects in terms of their position, including polar coordinates, three-dimensional Cartesian coordinates, vectors and limits.
MEAP Restriction: NOT CURRENTLY SCHEDULED FOR TESTING.

High School Benchmark 2: Locate and describe objects in terms of their orientation and relative position, including displacement vectors, phase shift, maxima, minima and inflection points; give precise mathematical descriptions of symmetries.
MEAP Restriction: NOT CURRENTLY SCHEDULED FOR TESTING.

High School Benchmark 3: Give precise mathematical descriptions of transformations and describe the effects of transformations on size, shape, position and orientation.
MEAP-Like Item: Students should engage in activities in which they give precise mathematical descriptions of transformations and their effects.

High School Benchmark 4: Describe the locus of a point by a rule or mathematical expression; trace the locus of a moving point.
MEAP Restriction: THIS BENCHMARK IS NOT SCHEDULED FOR TESTING AT THIS TIME.

High School Benchmark 5: Use concepts of position, direction and orientation to describe the physical world and to solve problems.
MEAP Restriction: Limited to what is testable within this standard. ONLY BENCHMARK 3 IS CURRENTLY SCHEDULED FOR TESTING. Students should engage in activities in which they give precise mathematical descriptions of transformations and describe the effects of transformation on size, shape, position, and orientation.

Content Standard 3: (Measurement)
Students compare attributes of two objects, or of one object with a standard unit, and analyze situations to determine what measurements should be made and to what level of precision.

High School Benchmark 1: Select and use appropriate tools; make accurate measurements using both metric and common units, and measure angles in degrees and radians.
MEAP Restrictions: Exclude radians from testing. Students should engage in activities in which they use measurement tools to accurately determine distance and angle measurements. They include measures computed from these and may require the use of trigonometric ratios.

High School Benchmark 2: Continue to make and apply measurements of length, mass, weight, time, temperature, area, volume, angle; classify objects according to their dimensions.
MEAP-Like Item: Students should engage in activities in which they accurately measure length, time, temperature, force, mass, area, volume, and angle, make computations with these measures, and use these measures to describe shapes.

High School Benchmark 3: Estimate measures with a specified degree of accuracy and evaluate measurements for accuracy, precision and tolerance.
MEAP-Like Item: Students should engage in activities in which they estimate linear measures and compute perimeter, area, and volume; recognize errors and find percent of error. Students may also be expected to recognize and use the concepts of accuracy and tolerance.

High School Benchmark 4: Interpret measurements and explain how changes in one measure may affect other measures.

High School Benchmark 5: Use proportional reasoning and indirect measurements, including applications of trigonometric ratios to measure inaccessible distances and to determine derived measures such as density.

MEAP-Like Item: Students should be engaged in activities in which they use trigonometric ratios or properties and relationships of shapes to make indirect measurements.

High School Benchmark 6: Apply measurement to describe the real world and to solve problems.
MEAP-Like Item: Students should engage in activities in which they solve geometric/measurement problems in real-world context.

III.   Data Analysis and Statistics

<u>**Content Standard 1**</u>: **(Collection, Organization and Presentation of Data)** Students collect and explore data, organize data into a useful form, and develop skill in representing and reading data displayed in different formats.

High School Benchmark 1: Collect and explore data through observation, measurement, surveys, sampling techniques and simulations.
MEAP Restriction: NOT CURRENTLY SCHEDULED FOR TESTING; schools should assess using performance tasks.

High School Benchmark 2: Organize data using tables, charts, graphs, spreadsheets and data bases.
MEAP-Like Item: Students should engage in activities in which they either construct or determine the appropriateness of several types of graphs and give meaning to the data using the graph as a tool. These activities should require them to organize data using tables, charts, graphs, spreadsheets and databases.

High School Benchmark 3: Present data using the most appropriate representation and give a rationale for their choice; show how certain representations may skew the data or bias the presentation.
MEAP-Like Item: Students should engage in activities in which they choose the most appropriate representation for given data and give justifications for not choosing alternative representations.

High School Benchmark 4: Identify what data are needed to answer a particular question or solve a given problem and design and implement strategies to obtain, organize and present those data.
MEAP-Like Item: Students should engage in activities in which they obtain, organize, and present data to solve problems involving real-world phenomena and to identify possible sources of bias.

<u>**Content Standard 2**</u>: **(Description and Interpretation)** Students examine data and describe characteristics of a distribution, relate data to the situation from which they arose, and use data to answer questions convincingly and persuasively.
High School Benchmark 1: Critically read data from tables, charts, or graphs and explain the source of the data and what the data represent.

MEAP-Like Item: Students should engage in activities in which they develop an understanding of data sources and answer questions regarding tables, charts and graphs.

High School Benchmark 2: Describe the shape of a data distribution and determine measures of central tendency, variability and correlation.
MEAP Restriction: Correlation will only be included if very limited. Graphing calculator appropriate. Students should engage in activities in which they determine a measure of central tendency and explore variability, correlation, and the shape of a data distribution.

High School Benchmark 3: Use data and their characteristics to draw and support conclusions.
MEAP-Like Item: Students should engage in activities in which they examine data and use the data to answer questions related to real world phenomena.

High School Benchmark 4: Critically question the sources of data; the techniques used to collect, organize and present data; the inferences drawn from the data; and the sources of bias and measures taken to eliminate such bias.
MEAP-Like Item: Students should engage in activities in which they critically evaluate the sources of data, methods of collection, organization, presentation, and conclusions drawn. Students should critically review for bias, mathematical errors, and logical inferences.

High School Benchmark 5: Formulate questions and problems and gather and interpret data to answer those questions.
MEAP Restriction: NOT CURRENTLY SCHEDULED FOR TESTING. Schools should assess using performance tasks. Students should engage in activities in which they demonstrate proficiency in formulating questions of interest, gathering appropriate data accurately, interpreting and critically analyzing solutions. Every phase of this process should include a critical review of possible sources of bias.

<u>**Content Standard 3:**</u> **(Inference and Prediction)**
Students draw defensible inference about unknown outcomes, make predictions, and identify the degree of confidence they have in their predictions.

High School Benchmark 1: Make and test hypotheses.
MEAP Restriction: Test using constructed-response items.

High School Benchmark 2: Design investigations to model and solve problems; also employ confidence intervals and curve fitting in analyzing the data.
MEAP Restriction: Exclude confidence intervals from testing. Curve fitting must be contextual.

High School Benchmark 3: Formulate and communicate arguments and conclusions based on data and evaluate their arguments and those of others.
MEAP Restriction: Test using Constructed Response items.

21

High School Benchmark 4: Make predictions and decisions based on data, including interpolations and extrapolations.

High School Benchmark 5: Employ investigations, mathematical models, and simulations to make inferences and predictions to answer questions and solve problems.
MEAP Restriction: Test using constructed response items.

IV. Number Sense and Numeration

<u>**Content Standard 1:**</u> **(Concepts and Properties of Numbers)**
Students experience counting and measuring activities to develop intuitive sense about numbers, develop understanding about properties of numbers, understand the need for and existence of different sets of numbers, and investigate properties of special numbers.

High School Benchmark 1: Develop an understanding of irrational, real and complex numbers.
MEAP Restriction: Exclude complex numbers from testing.
Students should engage in activities in which they recognize, use, and distinguish among various number systems with emphasis on irrational, real and complex numbers.
Complex numbers will not be tested on the current versions of MEAP tests.

High School Benchmark 2: Use the a+bi and polar forms of complex numbers
MEAP Restriction: NOT CURRENTLY SCHEULED FOR TESTING. Students should engage in activities in which they recognize, use and distinguish between rectangular and polar forms of complex numbers. Currently this benchmark is not scheduled for testing.

High School Benchmark 3: Develop an understanding of the properties of the real and complex number systems and of the properties of special numbers π, i, e. and conjugates.
MEAP Restriction: Exclude complex numbers and e from testing. Students should engage in activities in which they recognize and use the properties of real numbers. Complex numbers will not be tested on the current test.

High School Benchmark 4: Apply their understanding of number systems to model, and solve mathematical and applied problems.
MEAP Restriction: Limited to what is testable within this standard. That is, exclude complex numbers and e. Students should engage in activities in which they recognize and use the properties of various number systems to model and solve problems.

<u>**Content Standard 2:**</u> **(Representation and Uses of Numbers)**
Students recognize that numbers are used in different ways such as counting, measuring, ordering and estimating, understand and produce multiple representations of a number, and translate among equivalent representations.

High School Benchmark 1: Give decimal representations of rational and irrational numbers and coordinate and vector representations of complex numbers.

MEAP Restriction: Omit coordinate and vector representations of complex numbers from testing. Students should engage in activities in which they recognize and use various numbers, exponential notations, and graphic representations of rational, irrational, and complex numbers. Complex numbers will not be tested on the current version of the test.

High School Benchmark 2: Develop an understanding of more complex representations of numbers, including exponential and logarithmic expressions, and select an appropriate representation to facilitate problem solving.
MEAP Restriction: NOT CURRENTLY SCHEDULED FOR TESTING. [Usually covered after testing window. Students should engage in activities in which they extend exponential functions to include rational exponents and logarithmic functions with various bases. This benchmark is not currently scheduled for testing.

High School Benchmark 3: Determine when to use rational approximations and the exact values of numbers such as e, π and the irrationals.
MEAP Restriction: NOT CURRENTLY SCHEDULED FOR TESTING.

High School Benchmark 4: Apply estimation to increasingly complex situations.
MEAP-Like Item: Students should engage in activities in which they estimate the values of rational and irrational numbers.

High School Benchmark 5: Select appropriate representations for numbers, including representations of rational and irrational numbers and coordinate and vector representations of complex numbers in order to simplify and solve problems.
MEAP Restriction: Limited to what is testable within this standard. [Omit coordinate and vector representations of complex numbers. Students should engage in activities in which they select appropriate approximations to fit problem-solving situations.

Content Standard 3: (Number Relationships)
Students investigate relationships such as equality, inequality, inverses, factors and multiples, and represent and compare very large and very small numbers.

High School Benchmark 1: Compare and order real numbers and compare rational approximations to exact values.
MEAP-Like Item: Students should engage in activities in which they order real numbers and explain the relationship of approximations of irrational numbers to their exact values.

High School Benchmark 2: Express numerical comparisons as ratios and rates.
MEAP-Like Item: Students should engage in activities in which they use rates and ratios for comparisons of real numbers.

High School Benchmark 3: Extend the relationships of primes, factors, multiples and divisibility in an algebraic setting.
MEAP Restriction: NOT CURRENTLY SCHEDULED FOR TESTING.

High School Benchmark 4: Express number relationships using positive and negative rational exponents, logarithms and radicals.
MEAP restriction: NOT CURRENTLY SCHEDULED FOR TESTING.

High School Benchmark 5: Apply their understanding of number relationships in solving problems.
MEAP-Like Item: Students should engage in activities in which they solve problems using the testable concepts described in the other benchmarks of this standard.

V.   Numerical and Algebraic Operations and Analytical Thinking

Content Standard 1: (Operations and their Properties) Students understand and use various types of operations e.g., addition, subtraction, multiplication, division to solve problems.

High School Benchmark 1: Present and explain geometric and symbolic models for operations with real and complex numbers and algebraic expressions.
MEAP Restriction: Exclude complex numbers from testing.
Students should engage in activities in which they compute and explain operations such as, but not limited to, composition of algebraic expressions, basic operations with real and complex numbers, both symbolically and with models and drawings.
[NOTE: Area models similar to those below will appear in both constructed response and multiple choice items. Students will be required to express area as binomial factors or the resulting product of those factors. They will also be required to draw an area model given the area expressed in algebraic notation.]

High School Benchmark 2: Compute with real numbers, complex numbers, algebraic expressions, matrices and vectors using technology and, for simple instances, with paper-and-pencil algorithms.
MEAP Restriction: Exclude complex numbers and vectors from testing. Limit matrices to addition, subtraction and scalar multiplication.
Students should engage in activities in which they compute with real numbers and algebraic expressions. Scalar multiplication, vector operations, and geometric representation of operations with complex numbers, addition and subtraction of matrices are appropriate for testing

High School Benchmark 3: Describe the properties of operations with numbers, algebraic expressions, vectors and matrices, and make generalizations about the properties of given mathematical systems.
MEAP Restriction: Exclude vectors and matrices from testing in this round. [They may be included at a later time.]
Students should engage in activities in which they use and differentiate characteristics of numbers systems (e.g., whole numbers, natural numbers, rational and irrational real numbers, complex numbers). Characteristics might include properties (e.g., closure, associative, commutative, inverses, distributive).

High School Benchmark 4: Efficiently and accurately apply operations with real numbers, complex numbers, algebraic expressions, matrices and vectors in solving problems.
MEAP Restriction: Limited to what is testable within this standard (e.g., exclude complex numbers and vectors from testing and limit matrices to addition, subtraction and scalar multiplication). Graphic calculator appropriate.

Content Standard 2: (Algebraic and Analytic Thinking)
Students analyze problems to determine an appropriate process for solution, and use algebraic notations to model or represent problems.

High School Benchmark 1: Identify important variables in a context, symbolize them and express their relationships algebraically.
MEAP-Like Item: Students should engage in activities in which they translate problems in context into symbolic mathematical expressions and algebraic relationships using clearly defined variables.

High School Benchmark 2: Represent algebraic concepts and relationships with matrices, spreadsheets, diagrams, graphs, tables, physical models, vectors, equations and inequalities; and translate among the various representations.
MEAP Restriction: Exclude vectors and matrices from testing this round.

High School Benchmark 3: Solve linear equations and inequalities algebraically and non-linear equations using graphing, symbol-manipulating or spreadsheet technology; and solve linear and non-linear systems using appropriate methods.
MEAP Restriction: Limit non-linear systems to visual identification of intersection/using calculators. Graphing calculator appropriate.
Students should engage in activities in which they solve linear equations and linear inequalities, and linear systems and non-linear systems. Solution methods include algebraic techniques and use of appropriate technology.

High School Benchmark 4: Analyze problems that can be modeled by functions, determine strategies for solving the problems and evaluate the adequacy of the solutions in the context of the problems.
MEAP-Like Item: Students should engage in activities in which they model real life phenomena, analyze problems, and develop functions. They should make predictions based on their models, analyses, or functions.

High School Benchmark 5: Explore problems that reflect the contemporary uses of mathematics in significant contexts and use the power of technology and algebraic and analytic reasoning to experience the ways mathematics is used in society.
MEAP Restriction: Limit to what is testable within this standard; (i.e., exclude vectors and matrices, and limit non-linear systems to visual identification of intersection/use calculators). Graphing calculator appropriate. Students should engage in activities in

which they use algebraic and analytic reasoning and technology to solve problems in contemporary settings.

VI.   Probability and Discrete Mathematics

Content Standard 1: (Probability)
Students develop an understanding of the notion of certainty and of probability as a measure of the degree of likelihood that can be assigned to a given event based on the knowledge available, and make critical judgments about claims that are made in probabilistic situations.

High School Benchmark 1: Develop an understanding of randomness and chance variation and describe chance and certainty in the language of probability.
MEAP-Like Item: Students should engage in activities in which they extend their concepts of randomness, chance variation, and probability using real world phenomena. These activities allow students to become familiar with the notational and written language of probability.

High School Benchmark 2: Give a mathematical definition of probability and determine the probabilities of more complex events, and generate and interpret probability distributions.
MEAP Restriction: Not appropriate for testing in 2002, may be appropriate in the future. Students should engage in activities in which they make predictions and determine probabilities of complex events and probability distributions based upon real-world phenomena. Types of distributions include, but are not limited to, normal, binomial, uniform and Poisson (waiting time).

High School Benchmark 3: Analyze events to determine their dependence or independence and calculate probabilities of compound events.
MEAP-Like Item: Students should engage in activities in which they determine if two events are independent or dependent and determine the appropriate rules or procedures to calculate their probabilities.

High School Benchmark 4: Use sampling and simulations to determine empirical probabilities and, when appropriate, compare them to the corresponding theoretical probabilities; understand and apply the law of large numbers.
MEAP-Like Item: Students should engage in activities in which they use appropriate technology or manipulatives (spinners, dice, etc.) to show the relationship between experimental and theoretical probabilities.

High School Benchmark 5: Conduct probability experiments and simulations, to model and solve problems, including compound events.
MEAP Restriction: Do not use Monte Carlo method in testing. Students should engage in activities in which they use appropriate technology or manipulatives to model and solve problems involving probability of simple and compound events based on real-world phenomena.

Content Standard 2: (Discrete Mathematics)

Students investigate practical situations such as scheduling, routing, sequencing, networking, organizing and classifying, and analyze ideas like recurrence relations, induction, iteration, and algorithm design.

High School Benchmark 1: Derive and use formulas for calculating (the number of) permutations and combinations.
MEAP-Like Item: Students should be engaged in activities in which they have significant experience in making choices between the appropriate use of permutations or combinations in a real-world context. The activities should be extended in ways which require the student to use appropriate counting techniques, diagrams, and formulas.

High School Benchmark 2: Use sets and set relationships to represent algebraic and geometric concepts.
MEAP-Like Item: Students should engage in activities in which they represent algebraic and geometric concepts using sets and set relationships. These relationships include, but are not limited to, union, intersection, and complement.

High School Benchmark 3: Use vertex-edge graphs to solve network problems such as finding circuits, critical paths, minimum spanning tress and adjacency matrices.
MEAP Restriction: NOT APPROPRIATE FOR TESTING IN 2002. May be appropriate in the future.The students should be engaged in activities that require them to find circuits, critical paths, minimum spanning trees, and adjacency matrices using vertex-edge graphs.

High School Benchmark 4: Analyze and use discrete ideas, such as induction, iteration and recurrence relations.
MEAP-Like Item: Students should engage in activities in which they develop an understanding of iteration and recurrence relationships. These activities should be based in the counting of numbers and allow students to generalize patterns/sequences.

High School Benchmark 5: Describe and analyze efficient algorithms to accomplish a task or solve a problem in a variety of contexts, including practical, mathematical and computer-related situations.
MEAP Restriction: NOT APPROPRIATE FOR TESTING IN 2001. May be appropriate in the future. Students should engage in activities in which they describe and analyze efficient algorithms. Algorithms should provide a process to solve problems involving scheduling, routing, sequencing, networking, organizing and analyzing real-world phenomena.

High School Benchmark 6: Use discrete mathematics concepts as described above to model situations and solve problems; and look for whether or not there is a solution (existence problems), determine how many solutions there are (counting problems) and decide upon a best solution (optimization problems).

MEAP-Like Item: MEAP Restriction. NOT APPROPRIATE FOR TESTING IN 2001. May be appropriate in the future.

Students should engage in activities in which they model and solve existence, counting and optimization problems in a real-world context.

Maximum Effectiveness Of This Resource

The Michigan Curriculum Frameworks gives an overview of the curriculum that should be taught in every high school within the state of Michigan. A quick overview of this document reveals that it contains six strands, which contains 15 standards which contains over 70 benchmarks. Although the task of mastering a body of material this vast seems insurmountable at first glance, it becomes conquerable with the aid of technology.

A quick text search of the high school mathematics portion of the Michigan Curriculum Frameworks reveals that the terms

Data(27), Numbers(26), Problem(25), Solve(16) and Relationships(15)

appear more than any other. The repeated usage of these terms imply that certain solution strategies may be more effective than others. To maximize your effectiveness with the TI-83, we will focus on a limited number of calculator sequences. These calculator sequences include finding regression equations and programming.

For problems that involve formulas, students are encouraged to type the given formulas into their TI-83. Additional formulas and other calculator support may be found at www.education.ti.com or through a variety of search engines utilizing the search words "TI-83 programs".

MEAP HIGH SCHOOL MATHEMATICS TEST
Reference Sheet

Use this information as needed to answer questions on the Mathematics HST.

Miscellaneous

Distance = rate × time

Interest = principal × rate × time
Compound
Amount, $A = P(1 + r)^n$, where
P = principal, r = annual rate, n = time

Circumference of a circle = $\pi d = 2\pi r$

$$\pi = 3.14 = \frac{22}{7}$$

Algebra

Straight Line: $y = mx + b$

For points (x_1, y_1) and (x_2, y_2)

$$m = \frac{y_2 - y_1}{x_2 - x_1}$$

Quadratic Formula:
If $ax^2 + bx + c = 0$, $a \neq 0$, then

$$x = \frac{-b \pm \sqrt{b^2 - 4ac}}{2a}$$

Triangles

$$a^2 + b^2 = c^2$$

Examples:
$3^2 + 4^2 = 5^2$
$5^2 + 12^2 = 13^2$

Trigonometry

$$\sin A = \frac{\text{opposite}}{\text{hypotenuse}} = \frac{a}{c}$$

$$\cos A = \frac{\text{adjacent}}{\text{hypotenuse}} = \frac{b}{c}$$

$$\tan A = \frac{\text{opposite}}{\text{adjacent}} = \frac{a}{b}$$

Probability

$$_nC_r = C(n,r) = \frac{n!}{r!(n-r)!}$$

$$_nP_r = P(n,r) = \frac{n!}{(n-r)!}$$

$P(A \cup B) = P(A) + (P)B - P(A \cap B)$,
if A and B are two events

$P(A \cup B) = P(A) + P(B)$,
if A and B are mutually exclusive

Sequences/Series

Arithmetic Sequence: $a_n = a_1 + (n-1)d$

Geometric Series: $g_n = g_1 r^{n-1}$

$$S_n = \frac{g(1 - r^n)}{1 - r}$$

Area

◺	Triangle:	$A = \frac{1}{2}(\text{base}) \times \text{height}$
▭	Rectangle:	$A = \text{base} \times \text{height}$
⏢	Trapezoid:	$A = \frac{1}{2}(\text{sum of the bases}) \times \text{height}$
▱	Parallelogram:	$A = \text{base} \times \text{height}$
⊙	Circle:	$A = \pi r^2$
⬠	Regular Polygon:	$A = \frac{1}{2}(a) \times \text{perimeter}$

Reference Sheet Printed With Permission from, **"The Michigan Department of Education"**.

31

	Total Surface Area			Volume

Total Surface Area

Cone: SA = $\frac{1}{2}$(circumference of base) × (slant height) + πr^2

Pyramid: SA = (perimeter of base) × (slant height) + area of the base

Sphere: SA = $4\pi r^2$

Prism: SA = sum of the area of the faces

Cylinder: SA = circumference of the base × height + $2\pi r^2$

Cube: SA = 6 × (length of edge)^2

Volume

 V = $\frac{1}{3}\pi r^2$ × height

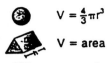 V = $\frac{1}{3}$area of the base × height

 V = $\frac{4}{3}\pi r^3$

V = area of the base × height

 V = πr^2 × height

V = (length of edge)^3

Trigonometry

Angle (Degrees)	Sin	Cos	Tan	Angle (Degrees)	Sin	Cos	Tan
0	.0000	1.0000	.0000	46	.7193	.6947	1.0355
1	.0175	.9998	.0175	47	.7314	.6820	1.0724
2	.0349	.9994	.0349	48	.7431	.6691	1.1106
3	.0523	.9986	.0524	49	.7547	.6561	1.1504
4	.0698	.9976	.0699	50	.7660	.6428	1.1918
5	.0872	.9962	.0875	51	.7771	.6293	1.2349
6	.1045	.9945	.1051	52	.7880	.6157	1.2799
7	.1219	.9925	.1228	53	.7986	.6018	1.3207
8	.1392	.9903	.1405	54	.8090	.5878	1.3764
9	.1564	.9877	.1584	55	.8192	.5736	1.4281
10	.1736	.9848	.1763	56	.8290	.5592	1.4826
11	.1908	.9816	.1944	57	.8387	.5446	1.5399
12	.2079	.9781	.2126	58	.8480	.5299	1.6003
13	.2250	.9744	.2309	59	.8572	.5150	1.6643
14	.2419	.9703	.2493	60	.8660	.5000	1.7321
15	.2588	.9659	.2679	61	.8746	.4848	1.8040
16	.2756	.9613	.2867	62	.8829	.4695	1.8807
17	.2924	.9563	.3057	63	.8910	.4540	1.9626
18	.3090	.9511	.3249	64	.8988	.4384	2.0503
19	.3256	.9455	.3443	65	.9063	.4226	2.1445
20	.3420	.9397	.3640	66	.9135	.4067	2.2460
21	.3584	.9336	.3839	67	.9205	.3907	2.3559
22	.3746	.9272	.4040	68	.9272	.3746	2.4751
23	.3907	.9205	.4245	69	.9336	.3584	2.6051
24	.4067	.9135	.4452	70	.9397	.3420	2.7475
25	.4226	.9063	.4663	71	.9455	.3256	2.9042
26	.4384	.8988	.4877	72	.9511	.3090	3.0777
27	.4540	.8910	.5095	73	.9563	.2924	3.2709
28	.4695	.8829	.5317	74	.9613	.2756	3.4874
29	.4848	.8746	.5543	75	.9659	.2588	3.7321
30	.5000	.8660	.5774	76	.9703	.2419	4.0108
31	.5150	.8572	.6009	77	.9744	.2250	4.3315
32	.5299	.8480	.6249	78	.9781	.2079	4.7046
33	.5446	.8387	.6494	79	.9816	.1908	5.1446
34	.5592	.8290	.6745	80	.9848	.1736	5.6713
35	.5736	.8192	.7002	81	.9877	.1564	6.3138
36	.5878	.8090	.7265	82	.9903	.1392	7.1154
37	.6018	.7986	.7536	83	.9925	.1219	8.1443
38	.6157	.7880	.7813	84	.9945	.1045	9.5144
39	.6293	.7771	.8098	85	.9962	.0872	11.430
40	.6428	.7660	.8391	86	.9976	.0698	14.301
41	.6561	.7547	.8693	87	.9986	.0523	19.081
42	.6691	.7431	.9004	88	.9994	.0349	28.636
43	.6820	.7314	.9325	89	.9998	.0175	57.285
44	.6947	.7193	.9657	90	1.0000	.0000	undefined
45	.7071	.7071	1.0000				

Reference Sheet Printed With Permission from, **"The Michigan Department of Education"**.

Mathematics Rubric for Open-ended Items

4 Points
- Carries appropriate strategy to correct solution.
- Understands the content of the problem
- Makes no meaningful errors.
- Provides clear and complete supporting calculations/arguments/justification.
- Uses available tools and representations correctly with correct results.

3 Points
- Uses an appropriate strategy.
- Shows significant understanding of the content of the problem.
- Makes minor errors (could complete the task with a non-instructional hint).
- Provides nearly complete supporting calculations/arguments/justification.
- Uses appropriate tools and representations correctly but reads the scale incorrectly or has an error in the interpretation

2 Points
- Begins with an appropriate strategy.
- Shows some understanding of the content of the problem.
- Makes some errors (requires some instruction prior to being able to complete the task).
- Provides some (but not complete) supporting calculations/arguments/justification.
- Recognizes an appropriate tool or representation but uses it inappropriately.

1 Point
- Attempts a strategy.
- Shows minimal understanding of the content.
- Makes serious errors but shows reasoning (requires significant instruction prior to being capable of completing the task).
- Provides few supporting calculations/arguments/justification.
- Uses inappropriate tools and/or representations.

0 Points
- Has no apparent strategy
- Shows no understanding of the content
- Shows no reasoning (significant instruction necessary prior to being capable of completing the task).
- Has no supporting calculations or explanations
- Uses no tools or representations.
- Leaves question blank or simply restates the problem.

Printed With Permission from The Michigan Department of Education

How Do I Type In (Input) A Formula Into My Calculator?

Start by pressing PRGM, once you are in program mode you will be able to type the following. After typing each line, press enter.

To get this on your screen	Type
ClrHome	PGRM, →(I/O), 8
Disp	PGRM, →(I/O), 3
Input	PGRM, →(I/O), 1
" "	Alpha, +
Any Letter	Alpha, The letter is above the number to the right
=	2nd, Math, 1
→	STO
Then, Stop, End or Else	PGRM, (Then enter the corresponding number or you can scroll down to the command and press enter).

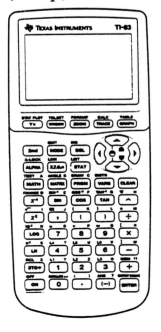

TI-83/TI-83 Plus Programs

(When stating your solutions, make sure to indicate all your steps and state "I used my TI-83")

Area of A Circle

Program:AREACIRC
ClrHome
Disp "Radius="
Input R
$\Pi R^2 \to A$
Disp "The Area="
Disp A

Area of a Parallelogram

Program:AREAPARA
ClrHome
Disp "BASE="
Input B
Disp "HEIGHT="
Input H
BH→A
Disp "THE AREA IS="
Disp A

Area of a Rectangle

Program:AREARECT
ClrHome
Disp "LENGTH="
Input L
Disp "WIDTH="
Input W
LW→A
Disp "THE AREA IS="
Disp A

Area of a Regular Polygon

Program:AREARPOL
ClrHome
Disp "REGULAR POLYGON"
Disp "NUMBER OF SIDES="
Input N
Disp "SIDE LENGTH"
Input L
Disp "APOTHEM"
Input R
.5LNR→A
Disp "THE AREA IS="
Disp A

Area of a Trapezoid

ClrHome
Disp "HEIGHT="
Input H
Disp "B1="
Input B
Disp "B2="
Input C
H(B+C)/2→A
Disp "AREA="
Disp A

Area of a Triangle

ClrHome
Disp "BASE="
Input B
Disp "HEIGHT="
Input H
.5BH→A
Disp "AREA="
Disp A

37

Circumference of a Circle
ClrHome
Disp "RADIUS="
Input R
$2\prod R \rightarrow C$
Disp "CIRCUMFERENCE="
Disp C

Combinations
ClrHome
Disp "TOTAL ELEMENTS="
Input N
Disp "NO.IN EACH SAMP"
Input R
$(N!)/(R!(N-R)!) \rightarrow C$
Disp "COMBINATIONS="
Disp C

Compound Interest
ClrHome
Disp "PRINCIPAL="
Input P
Disp "ANNUAL RATE="
Input R
Disp "YEARS="
Input Y
$P(1+R)^\wedge Y \rightarrow I$
Disp "COMPOUND INTR="
Disp I

Distance
PROGRAM:DISTANCE
ClrHome
Disp "RATE="
Input R
Disp "TIME="
Input T
$RT \rightarrow D$
Disp "DISTANCE="
Disp D

Permutations
ClrHome
Disp "TOTAL ELEMENTS="
Input N
Disp "NO.IN EACH SAMP"
Input R
$(N!)/(N-R)! \rightarrow P$
Disp "PERMUTATIONS="
Disp P

Quadratic Formula
ClrHome
Prompt A,B,C
$B^2 - 4AC \rightarrow D$
If D<0
Then
Disp " No Real Solutions"
Stop
End
If D = 0
Then
Disp " One Distinct Solution", -B/2A
Else
Disp "2 Solutions",
$(-B+\sqrt{(D)})/(2A)$, $(-B -\sqrt{(D)})/(2A)$
End

Simple Interest

```
PROGRAM:SIMPINTR
ClrHome
Disp "PRINCIPAL="
Input P
Disp "RATE="
Input R
Disp "TIME="
Input T
PRT→I
Disp "SIMPLE INTEREST="
Disp I
```

Surface Area of a Cone

```
Program:SACONE
ClrHome
Disp "RADIUS="
Input R
Disp "SLANT HEIGHT="
Input S
.5(2∏R)S+(∏R²)→A
Disp "SURFACE AREA="
Disp A
```

Surface Area of a Cube

```
PROGRAM:SACUBE
ClrHome
Disp "SIDE LENGTH"
Input L
6L²→A
Disp "SURFACE AREA="
Disp A
```

Surface Area of a Cylinder

```
PROGRAM;SACYLIND
ClrHome
Disp "RADIUS="
Input R
Disp "HEIGHT="
Input H
2∏RH+2∏R²→A
Disp "SURFACE AREA="
Disp A
```

Surface Area of a Prism

```
PROGRAM:SAPRISM
ClrHome
Disp "PERIM OF BASE="
Input P
Disp "HEIGHT="
Input H
Disp "AREA OF BASE="
Input B
PH+B→A
Disp "SURFACE AREA="
Disp A
```

Surface Area of a Pyramid

```
PROGRAM:SAPYRAMI
ClrHome
Disp "LENGTH="
Input L
Disp "WIDTH="
Input W
Disp "SLANT HEIGHT="
Input S
(2L+2W)S+LW→A
Disp "SURFACE AREA="
Disp A
```

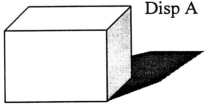

Surface Area of a Sphere
PROGRAM:SASPHERE
ClrHome
Disp "RADIUS="
Input R
$4\prod R^2 \to A$
Disp "SURFACE AREA="
Disp A

Slope of a Line
PROGRAM:SLOPE
ClrHome
Disp "(X1,Y1), (X2,Y2)"
Disp "X1="
Input A
Disp "Y1="
Input B
Disp "X2="
Input C
Disp "Y2="
Input D
$((B-D)/(A-C)) \to M$
Disp "SLOPE="
Disp M

Volume of a Cone
PROGRAM:VOLCONE
Disp "RADIUS'
Input R
Disp "HEIGHT"
Input H
$.333\prod R^2 H \to V$
Disp "THE VOLUME IS"
Disp V

Volume of a Cube
PROGRAM:VOLCUBE
Disp "SIDE LENGTH="
Input L
$L\wedge 3 \to V$
Disp "THE VOLUME IS="
Disp V

Volume of a Cylinder
Program:VOLCYLIN
ClrHome
Disp "Radius="
Input R
Disp Height="
Input H
$\prod R^2 H \to V$
Disp "THE VOLUME IS="
Disp V

Volume of a Prism
PROGRAM:VOLPRISM
ClrHome
Disp "B1"
Input B
Disp "B2"
Input C
Disp "H"
Input H
$.5BCH \to V$
Disp "THE VOLUME IS="
Disp V

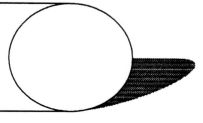

Volume of a Pyramid

```
PROGRAM:VOLPYRAM
ClrHome
Disp "BASE LENGTH="
Input L
Disp "BASE WIDTH="
Input W
Disp "PYRAMID HEIGHT=?"
Input H
.333LWH→V
Disp "THE VOLUME IS ="
Disp V
```

Volume of a Sphere

```
PROGRAM:VOLSPHERE
ClrHome
Disp "Radius="
Input R
1.333∏R^3→V
Disp "THE VOLUME IS="
Disp V
```

41

Sample Solution Strategies For The MEAP High School Mathematics Assessment Model April 2001

9. Pete currently earns $500 per month. He receives a 4% salary increase every three months. Three years from now, which amount will be closest in value to Pete's monthly salary?

A $560
B $630
C $800
D $1,000

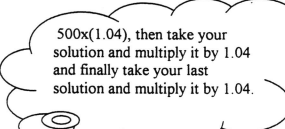

500x(1.04), then take your solution and multiply it by 1.04 and finally take your last solution and multiply it by 1.04.

13. The set x = {2, 5, 1, 9} represents the number of gallons of fuel used each day by a machine during the past four days. The number of gallons of fuel it uses on a given day determines the number of items the machine produces that day. The number if items produced each day is modeled by the following formula:

$$I(x) = x^2 - 14x + 45$$

Which value of x, used in the last four days, produced the most items?

A   1
B   2
C   5
D   9

Press Y =, (enter your function into Y1) and press table. Scroll until you find the corresponding values for 2, 5, 1 and 9.

43

19. List the following real numbers in order from greatest to least.

3Π   √69   33/8   8.14

A   3Π, √69, 8.14, 33/8

B   3Π, 8.14, √69, 33/8

C   3Π, √69, 33/8,   8.14

D   √69, 33/8, 8.14, 3Π

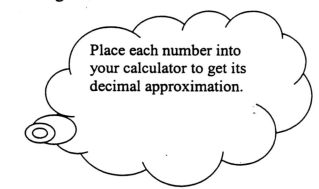

Place each number into your calculator to get its decimal approximation.

24. The table shows the speed of sound in air at different temperatures.
Speed of Sound at Different Temperatures

Air Temperature	Speed of Sound
16°C	1118 ft/sec
20°C	1126 ft/sec
24°C	1134 ft/sec

How far will sound travel in 20 seconds in air that is 29°C?

A   22,780 ft
B   22,840 ft
C   22,860 ft
D   22,880 ft

Press Stat, Enter, (place air data into L1, Speed data into L2), Stat, → (calc), 4, 2nd, 1, , (this is the comma button above the 7), 2nd, 2, enter. Place your equation into "Y=". Then press 2nd, graph, then scroll down to your x-value (air temperature) of 29. This is the speed for each second. Now multiply the answer by 20 to find the distance traveled in 20 seconds.

29. For the data table, determine the rule when x = n.

X	1	2	3	4	5
Y	25	20	15	10	5

A   25 + 5n
B   25 – 5n
C   30 + 5n
D   30 – 5n

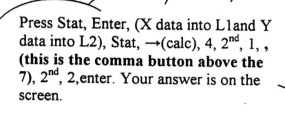

Press Stat, Enter, (X data into L1 and Y data into L2), Stat, →(calc), 4, 2nd, 1, , **(this is the comma button above the 7)**, 2nd, 2, enter. Your answer is on the screen.

Model Assessment Problems Printed With Permission from the
Michigan Department of Education

Interpreting Word Problems

The following list provides words that imply addition, subtraction, multiplication, and division.

Addition
Add
Sum
Plus
Both
Also
In all
Altogether
Together
Total
Joined
Combined

Subtraction
Left	Older
Less	Faster
Minus	Smaller
Difference	Wider
More	Greater
Off	Fewer
Change	Taller
Difference	Shorter
Profit	

Multiplication
Times
Product
Twice – Before The Question
Each– Before The Question
Per– Before The Question
Every– Before The Question
One– Before The Question

Division
Quotient
Equal amounts
Half
Thirds
Each – In The Question
Per– In The Question
Every– In The Question
One– In The Question

Michigan Curriculum Framework
Terminology

	Strand 1	*Strand 2*	*Strand 3*	*Strand 4*	*Strand 5*	*Strand 6*
1	Sequence	Loci	Data	Real Number	Numeric	Probability
2	Series	Graph	Survey	Rational Number	Numerical Equation	Certainty
3	Recursion	Distribution	Sample	Irrational Number	Numerical Value	Event
4	Matrix	Transformation	Simulation	Complex Number	Geometric	Random
5	Graph	Coordinate	Table	Integer	Symbolic	Variation
6	Vector	Synthetic	Chart	Natural Number	Algebraic	Event
7	Inference	Dimension	Graph	Equivalence Relation	Real Number	Probabilistic
8	Prediction	Isometry	Spreadsheet	Polar Coordinate	Rational Number	Dependence
9	Graphic Pattern	Coordinate	Database	Counting	Irrational Number	Independence
10	Numeric Pattern	Congruence	Skew	Measure	Complex number	Simulation
11	Linear Pattern	Similarity	Bias	Order	Integer	Empirical
12	Quadratic Pattern	Perpendicular	Distribution	Estimate	Natural Number	Theoretical
13	Family of Functions	Incidence	Mean	Equivalent	Equivalence Relation	Model
14	Rate	Polar Coordinates	Median	Exponential	Polar Coordinate	Discrete
15	Continuity	Vector	Mode	Logarithm	Matrix	Scheduling
16	Limit	Limit	Variability	Vector	Vector	Sequence
17	Distribution	Maximum	Correlation	Equality	Algorithm	Network
18	Deviation	Minimum	Inference	Inequality	Operation	Recurrence
19	Function	Inflection	Prediction	Inverse	Analytic	Induction
20	Linear	Symmetry	Hypothesis	Factor	Variable	Iteration
21	Linear Equation	Measurement	Modeling	Approximation	Spreadsheet	Algorithm

	Strand 1	*Strand 2*	*Strand 3*	*Strand 4*	*Strand 5*	*Strand 6*
22	Quadratic.	Precision	Confidence Interval	Ratio	Diagram	Permutation
23	Exponential	Metric	Interpolate	Rate	Table	Combination
24	Trigonometric Function	Angle	Extrapolate	Prime Number	Chart	Set
25	Composition of functions	Degree	Interval	Exponent	Graph	Vertex-Edge Graph
26	Piece-wise functions	Radian	Nominal		Model	Critical Paths
27	Recursively defined functions	Parallel	Ordinal		Equation	Minimum Spanning Trees
28	Inverse functions	Length	Ratio		Inequality	Adjacency Matrices
29		Mass	Data		Linear	Optimization Problems
30		Temperature	Survey		Nonlinear	
31		Area	Sample		Function	
32		Volume	Simulation		Reasoning	
33		Precision	Table			
34		Reasoning	Chart			
35		Indirect Measurements	Graph			
36		Sine	Spreadsheet			
37		Cosine	Database			
38			Skew			
39			Bias			
40			Distribution			

Strand 1:

Patterns, Relationships and Functions

Patterns – Students recognize similarities and generalize patterns to create models and make predictions, describe the nature of patterns and relationships, and construct representations of mathematical relationships.

Patterns, Relationships and Functions

1.Patterns – Students recognize similarities and generalize patterns to create models and make predictions, describe the nature of patterns and relationships, and construct representations of mathematical relationships.

- Analyze and generalize mathematical patterns including sequences, series and recursive patterns.
- Analyze, interpret and translate among representations of patterns including tables, charts, graphs, matrices and vectors.
- Study and employ mathematical models of patterns to make inferences, predictions and decisions.
- Explore patterns (graphic, numeric, etc.) characteristics of families of functions; explore structural patterns within systems of objects, operations or relations.
- Use patterns and reasoning to solve problems and explore new content.

2.Variability and Change – Students describe the relationships among variables, predict what will happen to one variable as another variable is changed, analyze natural variation and sources of variability, and compare patterns of change.

- Identify and describe the nature of change and begin to use the more formal language such as rate of change, continuity, limit, distribution and deviation.
- Develop a mathematical concept of function and recognize that functions display characteristic patterns of change (e.g., linear, quadratic, exponential)

Strand 1: Patterns, Relationships and Functions

- Expand their understanding of function to include non-linear functions, inverses of functions, and piece wise and recursively-defined functions.
- Represent functions using symbolism such as matrices, vectors, and functional representation ($f(x)$).
- Differentiate and analyze classes of functions including linear, power, quadratic, exponential, circular and trigonometric functions and realize that many different situations can be modeled by a particular type of function.
- Increase their use of functions and mathematical models to solve problems in context.

Strand 1 Terms

Patterns, Relationships and Functions

Sequence
Series
Recursion

Matrix

Graph
Vector
Inference
Prediction

Graphic Pattern

Numeric Pattern
Linear Pattern
Quadratic Pattern
Family of Functions
Rate
Continuity
Limit

Distribution

Deviation

Function

Linear

Linear equation

Quadratic.

Exponential

Trigonometric Function
Composition of functions
Piece wise functions
Recursively defined functions
Inverse functions

Strand 1: Patterns, Relationships and Functions

Patterns, Relationships and Functions

Lesson 1.1

The senior class at Jefferson High School has decided to have a conservation project. They will plant red and silver Maple Trees. The following chart represents the total number of trees planted at the end of each week. Use the information given to answer the following questions:

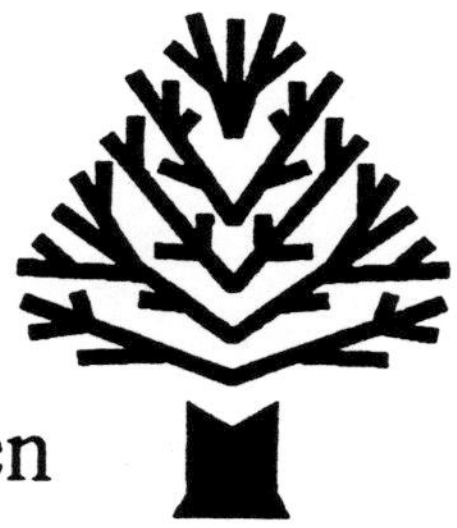

Week	Red Maples	Silver Maples
1	150	50
2	230	125
3	310	250
4	390	425
5	470	650

1. Describe the pattern of change in the number of Red Maple trees planted.

a) Constant
b) Linear
c) Quadratic
d) Cubic

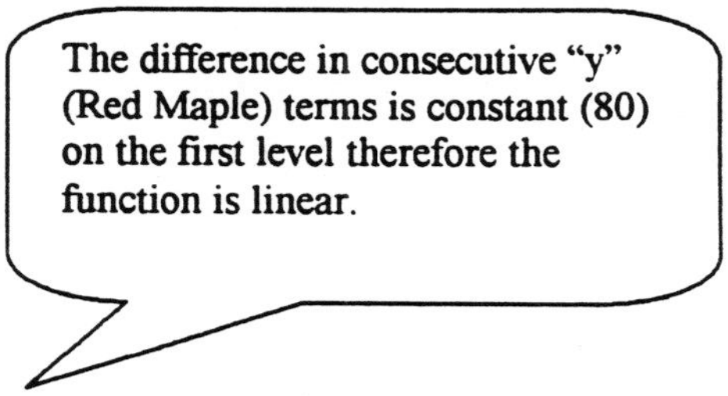

The difference in consecutive "y" (Red Maple) terms is constant (80) on the first level therefore the function is linear.

2. Describe the pattern of change in the number of Silver Maples planted.

   a) Constant
   b) Linear
   c) Quadratic
   d) Cubic

   > The difference in consecutive "y" (Silver Maple) terms is not constant on the first level, but a constant is seen on the second level (50) therefore the function is quadratic.

3. Assuming the planting rate continues, determine the number of Red Maple Trees that will be planted at the end of week 12.

   a) 1800
   b) 1200
   c) 1030
   d) 950

   > A pattern appears to be developing...The difference in consecutive "y" (Red Maple) terms is constant (80), therefore just extend the pattern until you reach the 12th week.

4. Assuming the planting rate continues, determine the number of Silver Maple Trees that will be planted at the end of week 12.

   a) 14465
   b) 12165
   c) 3860
   d) 3625

   > The difference in consecutive "y" (Silver Maple) terms is not constant on the first level, but a constant is seen on the second level (50) therefore just extend the pattern until you reach the 12th week.

5. Find the equation that best fits the relationship between the week and number of Red Maple Trees planted. Fully explain your work.

> Press Stat, Enter, (place week data into L1, Red Maple data into L2), Stat, →(calc), 4, 2nd, 1, , **(this is the comma button above the 7)**, 2nd, 2,enter.

6. Find the equation that best fits the relationship between the week and the number of Silver Maple Trees planted.
   Fully explain your work.

> Press Stat, Enter, (place week data into L1, Silver Maple data into L3, Stat, →(calc), 5, 2nd, 1, , , 2nd, 3,enter.

7. Estimate when the number of Red and Silver Maples Trees planted is the same. Fully explain how you arrived at your result.

To find when two entities are the same, you must set them equal to each other. You want to know when is $25x^2+25=80x+70$ or, if you move everything to the left side of the equation, when is $25x^2-80x-45=0$. You will need your quadratic formula program to solve this equation. Make sure you have programmed the formula into your calculator from the formula section of this book. Your Math teacher may also have several programs.

If you use our program, input the following: PRGM, Quad Formula A=25,enter, B=-80, enter, C=-45, enter. The answers are given as (3.69 and -.488). Since we are dealing with time (weeks) you can disregard the negative answer. The solution is 3.69 weeks or between the 3rd and 4th week.

8. The senior class at Jefferson High School had a successful fundraiser. They sold 322 roses to the student body. The following information gives a distribution of the roses sold.

Freshmen 55 roses
Sophomores 83 roses
Juniors 112 roses
Seniors 72 roses

Represent the distribution of roses sold utilizing the following:
a) Bar Graph
b) Pie Chart
c) Table

Show all calculations.

For a bar graph and table, you can use the actual numbers but for a pie chart, you must convert your values to percents. Remember for percents you need part/total.

Strand 1: Patterns, Relationships and Functions

Lesson 1.2

Paul's Expenditures

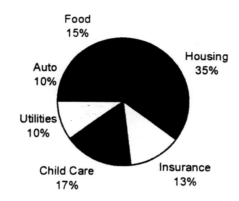

Food
15%

Housing
35%

Auto
10%

Utilities
10%

Child Care
17%

Insurance
13%

---

1. If Paul earned $2,732.18, how much did he spend on food and utilities?

> When multiplying with percents remember to move your decimal two(2) places to the left then multiply.

a) $725.24
b) $683.05
c) $409.83
d) $273.22

2. If Paul earned $2,732.18, how much more did he spend on Housing than Child Care?

a) $956.20
b) $528.73
c) $491.79
d) $464.47

3. Food and utilities comprise what fraction of Paul's expenses?

a) about one twenty-fifth
b) about one fifth
c) about one fourth
d) about one third

4. Paul's Financial Planner recommends that he saves more money. If Paul saves money by reducing his child care expenses to 12% and utilities to 7% how much additional money would he save?

a) $737.69
b) $464.47
c) $273.22
d) $218.57

Examine the Geometric pattern below. Each ☐ represents a square unit.

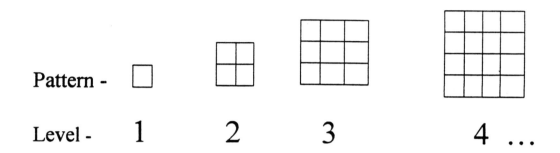

Pattern -

Level -   1   2   3   4 ...

5.Convert the information above into chart form. Fully explain your work.

Level	Length	Width	Perimeter	Area
1				
2				
3				
4				

6.Utilizing the chart, find the equation that best fits the relationship between Length and Perimeter. Use this equation to find the perimeter of the object with length 10 units. Fully explain your work.

Was the equation generated for perimeter constant, linear, quadratic or cubic? Explain how your response based upon the concept of perimeter.

Check for finite differences!

7. Utilizing the chart, find the equation that best fits the relationship between the Length and Area. Use this equation to find the Area of the object with a length of 10 units. Fully explain your work.

Was the equation generated for Area constant, linear, quadratic or cubic? Explain why you chose your response based upon the concept of area.

Check for finite differences!

8. The two equations that you generated for Perimeter and Area apply to which geometric shape? Find the Perimeter and Area for that shape when the length is 12 units. Fully explain your work.

Lesson 1.3

The following rates represent cellular calling plans for ABC Cellular Company.

Plan A- $20 monthly fee plus $0.12 per minute
Plan B- $25 monthly fee plus $0.10 per minute
(per minute charge represents anytime/anywhere minutes)

1.At how many minutes is the cost for plans A & B the same?

a) 250 minutes
b) 225 minutes
c) 200 minutes
d) 175 minutes

2.Rochelle chose Plan A for her cellular calling plan. Last month she talked for 932 minutes. How much would she have saved if she had chosen Plan B?

a) $131.84
b) $118.20
c) $19.84
d) $13.64

3.Rochelle's cellular bill this month is $171.56. If she is on Plan A, how many minutes did she talk?

a) 1466 minutes
b) 1263 minutes
c) 1087 minutes
d) 1032 minutes

Strand 1: Patterns, Relationships and Functions

	Ages	Cost	
Wendy's Admission	0-6	FREE	**Water World Rates**
	7-12	(.10 x Age)	
	13-17	(.25 x Age)	
	18 and over	$5.00	

4. The Brown family has decided to spend the day at Wendy's Water World. The Brown's have a 5year old son, an 11year old son and a set of 15year old twin daughters. How much will Mr. Brown pay for two adults (over 18 years of age) and their four children?

a) $18.60
b) $17.50
c) $14.85
d) $13.60

5. The Brown family decided to return to Wendy's Water World the following week. Mr. and Mrs. Brown brought their children ages 5, 11, 15 and 15 years and two cousins ages 14 and 16 years. How much more did Mr. Brown pay for the adult tickets than the children's (under 18 years) tickets?

a) He paid more for the children's tickets
b) $18.60
c) $10.00
d) $ 6.10

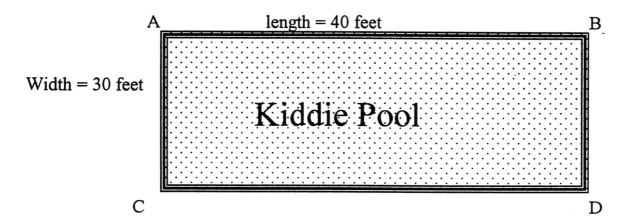

A   length = 40 feet   B

Width = 30 feet

Kiddie Pool

C   D

6. Micheal is a lifeguard at the Kiddie Pool. The lifeguard booth is stationed at point A and there is a stray flipper located at point D. If the pool is rectangular, what is the shortest distance from point A to point D?

a) 70 feet
b) 60 feet
c) 50 feet
d) 40 feet

Pythagorean Theorem!

Strand 1: Patterns, Relationships and Functions

7. Some parents have complained that the Kiddie Pool is too small, so the owners have decided to increase the pool's length 50% and it's width 100%. These changes will increase the Area and Perimeter by what percent? Fully explain your work.

8. With the new renovations complete, Kiddie Pool decided to raise it's admission rates. Last year, with the old pool, Kiddie Pool generated an average of $7.50 per square foot each day. With a 50% increase in length and 100% increase in width, estimate the amount Kiddie Pool should generate daily if their amount generated per square foot remains the same. Fully explain your work.

Strand 2:
Geometry and Measurement

Students use analytical and spatial concepts of shape, size, position, measurement and dimension to understand and interpret the three-dimensional world in which we live.

73

Geometry and Measurement

3.Shape and Shape Relationships –
Students develop spatial sense, use shape as an analytic and descriptive tool, identify characteristics and define shape, identify properties and describe relationships among shapes.

- Use shape to identify plane and solid figures, graphs, loci, functions, and data distributions.
- Determine necessary and sufficient conditions for the existence of a particular shape and apply those conditions to analyze shapes.
- Use transformational, coordinate or synthetic methods to verify (prove) the generalizations they have made about properties of classes of shapes.
- Draw and construct shapes in two and three dimensions and analyze and justify the steps of their constructions.
- Study transformations of shapes using isometries, size transformations and coordinate mappings.
- Compare and analyze shapes and formally establish the relationships among them, including congruence, similarity, parallelism, perpendicularity and incidence.

4.Position – Students identify locations of objects, identify location relative to other objects, and describe the effects of transformations (e.g., sliding, flipping, turning, enlarging, reducing) on an object.

- Locate and describe objects in terms of their position, including polar coordinates, three dimension Cartesian coordinates, vectors, and limits.

75

- Locate and describe objects in terms of their orientation and relative position, including displacement (vectors), phase shift, maxima, minima and inflection points; give precise mathematical descriptions of symmetries.
- Give precise mathematical descriptions of transformations and describe the effects of transformations on size, shape, position and orientation.
- Describe the locus of a point by a rule or mathematical expression; trace the locus of a moving point.
- Use concepts of position, direction and orientation to describe the physical world and to solve problems.

5. Measurement – Students compare attributes of two objects or of one object with a standard (unit), and analyze situations to determine what measurement(s) should be made and to what level of precision.

- Select and use appropriate tools; make accurate measurements using both metric and common units, and measure angles in degrees and radians.
- Continue to make and apply measurements of length, mass (weight), time, temperature, area, volume, angle; classify objects according to their dimensions.
- Estimate measures within a specified degree of accuracy and evaluate measurements for accuracy, precision and tolerance.
- Interpret measurements and explain how changes in one measurement may affect other measures.
- Use proportional reasoning and indirect measurements, including applications of trigonometric ratios, to measure inaccessible distances and to determine derived measures such as density.

Strand 2 Terms
Geometry and Measurement

Loci
Graph
Distribution
Transformation
Coordinate
Synthetic
Dimension

Isometry
Coordinate

Congruence

Similarity

Perpendicular

Incidence

Polar Coordinates

Vector

Limit

Maximum

Minimum

Inflection

Symmetry

Measurement

Precision

Metric

Angle

Degree

Radian
Parallel
Length
Mass
Temperature
Area
Volume
Precision
Reasoning
Indirect Measurements
Sine
Cosine
Tangent

Lesson 2.1

1. The Voting Commission reported that in the last election, there was strong representation from 18 and 60 year olds. The following distribution gives the age of voters versus the number who voted. The distribution is best described by what type of function ?

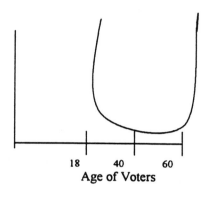

Age of Voters

a) Linear
b) Quadratic
c) Cubic
d) Exponential

2. If the diagonals of a parallelogram are perpendicular and congruent, then the parallelogram is a

a) Kite
b) Rectangle
c) Rhombus
d) Square

3. A building casts a 45° angle with the ground. Describe the triangular shadow.

a) Equiangular
b) Equilateral
c) Isosceles
d) Obtuse

4. Kim combined 2 quarts of punch with two cups of lemonade. How many more ounces of punch are there in the mixture than lemonade?

a) 80 ounces
b) 72 ounces
c) 64 ounces
d) 48 ounces

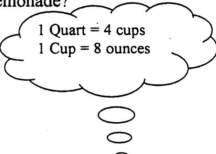

1 Quart = 4 cups
1 Cup = 8 ounces

5.

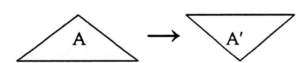

The mapping of A onto A′ represents which of the following

a) Glide
b) Reflection
c) Rotation
d) Translation

6.Tammy's circular ring has a radius of 7mm. What is the circumference of the ring?
Use π = 22/7

a) 308mm
b) 154mm
c) 88mm
d) 44mm

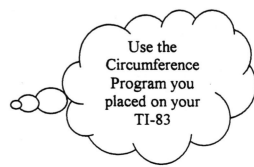

Use the Circumference Program you placed on your TI-83

7. To complete his remodeling jobs, Robert needs 1 yard of steel piping. If Dave gives him 2 feet of piping, how many more feet of piping does he need?

a) 12 feet
b) 3 feet
c) 1 foot
d) ½ foot

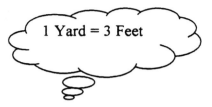

1 Yard = 3 Feet

8. Pluto's Plumbing charges $55.00 per hour plus a $100.00 service fee. Express Pluto's charges as a function and graph the function.

Pluto has been in business for 50 years and plans to celebrate the event by having a sale. They will reduce their service fee by 50%. How will a 50% reduction in the service fee affect the graph of the original function? Fully explain your work.

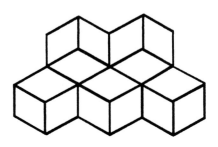

Lesson 2.2

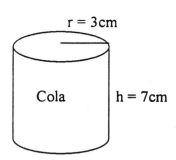

1.Karen's cola can has a height of 7cm. If its radius is 3cm, what is the volume of Karen's can? Use $\pi = 22/7$

a) 264cm³
b) 198cm³
c) 132cm³
d) 66cm³

Use the Volume of a Cylinder Program you placed on your TI-83

2.What is the greatest precision that can be achieved by this clock?

a) nearest 1 minute
b) nearest 5 minutes
c) nearest 9 minutes
d) nearest 15 minutes

3. If the diagonals of a parallelogram are congruent, then the parallelogram is

a) A Kite
b) A Rectangle
c) A Regular Pentagon
d) A Rhombus

4. David wants to determine how much water his container will hold. Which unit of measure will allow him to measure this quantity?

a) Area
b) Circumference
c) Density
d) Volume

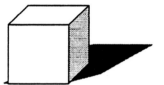

5. Freddie is using styrofoam balls in his Science Fair project. He plans to paint a large ball red to represent Jupiter. What unit of measure will allow him to measure the amount of space covered by the paint?

a) Area
b) Circumference
c) Density
d) Surface Area

6. Which unit would allow you to most precisely measure the length of your index finger?

a) millimeter
b) centimeter
c) meter
d) kilometer

7.

Triangle Park

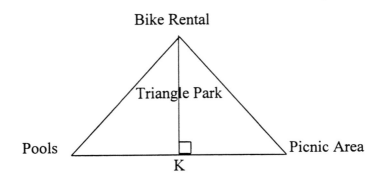

Bike Rental

Triangle Park

Pools

Picnic Area

K

Triangle Park is an equiangular shaped park located on the outskirts of Iona City. If the point "K" is halfway between the Pools and Picnic Area and is 4 miles from the Picnic Area, what is the shortest distance from the Bike Rental to the Pools?

a) 16 miles
b) 12 miles
c) 10 miles
d) 8 miles

Review the definition of equiangular. What does it say about the lengths of the sides?

8. Ricky must calculate the surface area and volume of the cylinder that he will use in his Chemistry experiment. If the cylinder has a height of 14cm and a diameter of 4cm, what is its surface area and volume? Fully explain your work.

Use the Volume & Surface Area programs for a cylinder that you placed on your TI-83

Lesson 2.3

1. What is the greatest degree of precision that can be achieved by this clock?

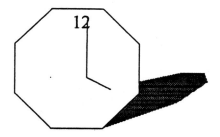

a) nearest 60 minutes
b) nearest 30 minutes
c) nearest 12 minutes
d) nearest 1 minute

2. Tina must frost Betty's birthday cake. If the cake has a cylindrical shape with a height of 7 inches and a radius of 10 inches, what is its surface area? (Yes, she will frost the bottom!) Use $\pi = 22/7$

a) 2200 in²
b) 1068 in²
c) 440 in²
d) 140 in²
e) 70 in²

Use the Surface Area Program you placed on your TI-83

3. Jackie needs 130 ounces of cola to make her famous cola float for the homecoming dance. If she already has 3 quarts of cola, how many more ounces of cola does Jackie need?

a) 162 ounces
b) 96 ounces
c) 34 ounces
d) 32 ounces

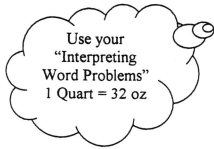

Use your "Interpreting Word Problems" 1 Quart = 32 oz

Strand 2: Geometry and Measurement

4.Mrs.Johnson's Home Economics class prepared 4 foot long hot dogs. How many 1 inch appetizers can be made from the 4 foot long hot dogs?

a) 12
b) 16
c) 48
d) 96

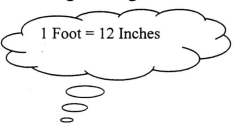

1 Foot = 12 Inches

5.The Plastic Shop ordered a shipment of cubical containers. If a container has a side length of 7 inches, what is its volume?

a) 343 cm³
b) 294 cm³
c) 98 cm³
d) 49 cm³

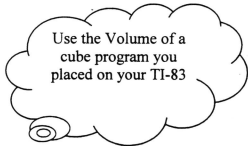

Use the Volume of a cube program you placed on your TI-83

6.Anita spent 1 hour 15 minutes washing dishes, 37 minutes ironing and 13 minutes sweeping. Altogether, how many minutes did Anita spend working?

a) 165 minutes
b) 150 minutes
c) 135 minutes
d) 125 minutes

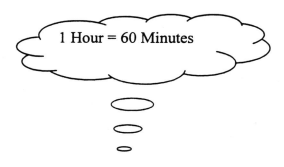

1 Hour = 60 Minutes

7.

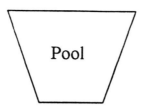

Pool

The King's own a trapezoidal shaped swimming pool. If the lengths of its bases are 8 feet and 14 feet respectively with a height of 20 feet, what is its area?

a) 224 ft²
b) 220 ft²
c) 112 ft²
d) 66 ft²

> Use the Area of a Trapezoid Program you placed on your TI-83

8. The ACE Architectural Firm constructed a 5inch by 8inch rectangular scale drawing of the Brown's dining room. If 3 feet = 1 inch, what is the area of the actual dining room? Fully explain your work.

> Don't forget to multiply the scale factor by all dimensions!

Strand 3:

Data Analysis and Statistics

Students organize, interpret and transform data into useful knowledge to make predictions and decisions based on data.

Data Analysis and Statistics

6.Collection, Organization and Presentation of Data – Students collect and explore data, organize data into a useful form, and develop skill in representing and reading data displayed in different formats.

- Collect and explore data through observation, measurement, surveys, sampling techniques and simulations.
- Organize data using tables, charts, graphs, spreadsheets and data bases.
- Present data using the most appropriate representation and give a rationale for their choice; show how certain representations may skew the data or bias the presentation.
- Identify what data are needed to answer a particular question or solve a given problem, and design and implement strategies to obtain, organize and present those data.

7.Description and Interpretation – Students examine data and describe characteristics of a distribution, relate data to the situation in which they arose, and use data to answer questions convincingly and persuasively.

- Critically read date from tables, charts or graphs and explain the source of the data and what the data represent.
- Describe the shape of a data distribution and determine the measures of central tendency, variability and correlation.
- Use the data and their characteristics to draw and support conclusions.
- Critically question the sources of data; the techniques used to collect, organize and present data; the inferences drawn from the data; and the sources of bias and measures taken to eliminate such bias.
- Formulate questions and problems, and interpret data to answer those questions.

Strand 3: Data Analysis and Statistics

8. Inference and Prediction – Students draw defensible inferences about unknown outcomes, make predictions, and identify the degree of confidence they have in their predictions.

- Make and test hypothesis.
- Design investigations to model and solve problems; also employ confidence intervals and curve fitting in analyzing the data.
- Formulate and communicate arguments and conclusions based on data and evaluate their arguments and those of others.
- Make predictions and decisions based on data, including interpolations and extrapolations.
- Employ investigations, mathematical models and simulations to make inferences and predictions to answer questions and solve problems.

Strand 3 Terms

Data Analysis and Statistics

Data
Survey
Sample
Simulation
Table
Chart
Graph
Spreadsheet
Database
Skew
Bias
Distribution
Mean
Median
Mode
Variability
Correlation
Inference
Prediction
Hypothesis
Modeling
Confidence Interval
Interpolate
Extrapolate
Interval
Nominal
Ordinal
Ratio

Lesson 3.1

1. Janie earned the following scores in her Algebra class 76, 84, 93, and 67. What is her mean score?

a) 75
b) 76
c) 78
d) 80

Mean is the average... DO NOT make the mistake of dividing by 2. Since there are 4 numbers, you will divide by 4.

2. Tom has 3 nickels, 4 quarters, 2 dimes, and 3 pennies in his pocket. If Tom selects 1 coin randomly, what is the probability that it will be a penny?

a) 1/6
b) 1/4
c) 1/3
d) 1/2
e) 2/3

Remember...(Part/Total) – Which part (or how many pennies divided by the total number of coins). Do not forget to reduce your fraction.

3. How many outfits (1 top and 1 bottom) can Sam create from 5 shirts and 6 pair of slacks?

a) 11
b) 25
c) 30
d) 36

With multiple combinations, multiply combinations

4.How many three digit numbers can be formed using the digits 1, 2, & 3? (No number can be repeated)

a) 12
b) 9
c) 6
d) 4
e) 3

5.If two coins are flipped at the same time, what is the probability that 2 heads will appear?

a) 1
b) 1/2
c) 1/4
d) 1/6

The chart shows comparable prices of mufflers at various muffler stores.

Business	Cost
Dave's Mufflers	$25
Sam's Shoppe	$28
Mr. Muff	$32
Cars-R-Us	$28
Car Muff	$30

6.What is the median muffler cost?

a) $28
b) $29
c) $30
d) $32

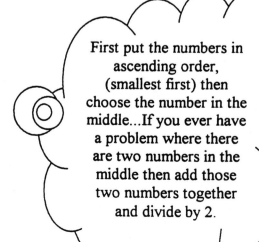

First put the numbers in ascending order, (smallest first) then choose the number in the middle...If you ever have a problem where there are two numbers in the middle then add those two numbers together and divide by 2.

7.Cornell has a bag of 70 marbles. There are 15 red, 30 green, 7 blue, 8 gray, 6 yellow, and 4 clear marbles. If Cornell reaches into his bag without looking, what is the probability that he would draw a gray marble?

a) 3/35
b) 1/10
c) 4/35
d) 3/14

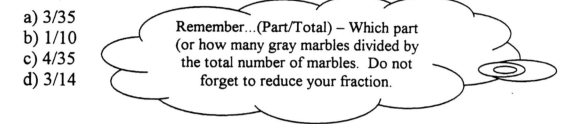

Remember...(Part/Total) – Which part (or how many gray marbles divided by the total number of marbles. Do not forget to reduce your fraction.

8. Dave and Carl grossed $162 mowing lawns. Twenty dollars was used to purchase lawnmower supplies, $12 for gasoline and $15 for trash bags. The remaining money was split evenly between Dave and Carl. Construct a pie chart to represent the distribution of the $162. Show all your work..

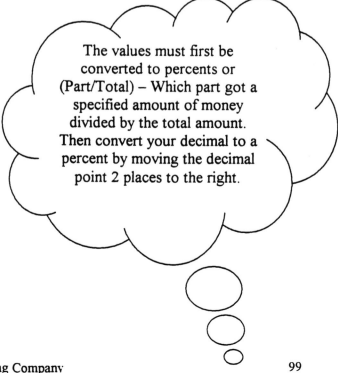

The values must first be converted to percents or (Part/Total) – Which part got a specified amount of money divided by the total amount. Then convert your decimal to a percent by moving the decimal point 2 places to the right.

Lesson 3.2

1. How many meals can be created from 3 meats, 4 vegetables, and 5 desserts? (A meal consists of 1 meat, 1 vegetable, and 1 dessert)

a) 12
b) 20
c) 30
d) 60

2. During the month of March, Tina lost 4lbs the first week, 2lbs the second week, 3lbs the third week, and 3lbs the fourth week. What was Tina's mode weight loss?

a) 12lbs
b) 5lbs
c) 4.5lbs
d) 3lbs

3. Sherry is shopping for a new movie camera. She has priced them at $3500 and $4100. What is the mean cost?

a) $9400
b) $7600
c) $3800
d) $600

4. A locker contains 3 math books, 4 English books, and 2 history books. If one book is randomly selected, what is the probability that it will be a math book?

a) 9
b) 6
c) 4/9
d) 1/3

English1 Semester Grades
(In Percents)

	Quiz1	Quiz2	*Test1*	Quiz3	Quiz4	*Test2*
Ann	72	83	*46*	95	62	*75*
Bea	59	75	*72*	79	86	*84*
Carl	52	64	*75*	83	91	*93*
David	85	79	*81*	82	88	*91*
Elam	85	64	*97*	75	78	*80*

5. Who had the highest overall test average?

a) Bea
b) Carl
c) David
d) Elam

6. If the tests are worth 70% of the grade and quizzes are worth 30%, who had the highest overall grade?

a) Bea
b) Carl
c) David
d) Elam

7. The English1 semester grades are

a. Interval
b. Nominal
c. Ordinal
d. Ratio

8. The Drivers Training final exam scores were 54, 80, 84, 70, 65, 91, 78, 0, 12, and 72.

- Organize the data in a stem and leaf plot.
- Would the mean or median be more representative of the average test score? Explain and support your answer.

Lesson 3.3

1. Rochelle is attempting to coordinate her hats and shoes. If she has 6 hats and 8 pairs of shoes, how many different combinations can she create?

a) 48
b) 24
c) 18
d) 12

2. If a coin and a die are tossed, what is the probability that a number greater than 4 would appear on the die and tails would appear on the coin?

a) 1/2
b) 1/4
c) 1/6
d) 1/8

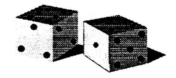

3. In Mrs. Jones Algebra class, there are 4 sophomores, 8 juniors, and 8 seniors. What is the probability that a student chosen at random will NOT be a senior?

a) 3/5
b) 2/5
c) 1/16
d) 1/36

4. Chris paid $25 for a pair of tennis shoes, $32 for a pair of dress shoes, and $2 for a pair of shower shoes. What was the mean amount Chris spent on shoes?

a) $19
b) $19.67
c) $25
d) $29.50

5.During a recent poll, 7out of 17 people surveyed preferred candy to fruit. If 130 people were polled, what reasonable conclusion can be drawn?

a) They must have polled kids
b) Everyone should brush between snacks
c) Over half of them prefer candy to fruit
d) Less than half prefer candy to fruit

Meyers High School

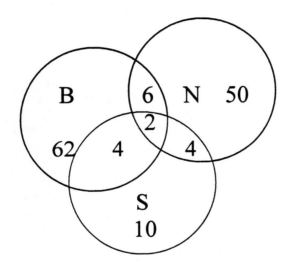

B=Students in the Band
N=Students in the National Honor Society
S=Students in the Student Council
U=529 students at Meyers High School

6.About what percentage of the students at Meyers High School are in the National Honor Society?

a) Between 40% and 50%
b) Between 30% and 38%
c) Between 20% and 28%
d) Between 10% and 18%

7. The Yearbook Class wants to poll students to determine the most memorable event of the school year. Which sampling would give the least bias?

a) Survey the Football Team
b) Survey the Senior Class
c) Survey the Journalism Staff
d) Survey 50 students at random

Student	Humerus Length (cm)	Height (cm)
A	55	239
B	57	245
C	59	251
D	61	257
E	63	263

8. Forensic scientists can estimate a male's height by measuring the length of his Humerus (upper arm bone). Paul decided to find an equation that would best fit the male's height in his Biology Lab. Of the 5 males, the above chart represents their Humerus lengths and heights:

Press Stat, Enter, (place Humerus data into L1, Height data into L2) Stat, →(calc), 4, 2nd, 1, , **(this is the comma button above the 7)**, 2nd, 2, enter.

a. Graph the Humerus length vs. the Height
b. Determine the equation of the line that best fits your data.

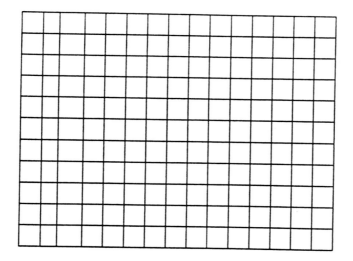

c. What is the average height of the males?

d. Using the equation you generated in Part (b), determine the height of a male who has a Humerus length of 48cm.

Strand 4:

Number Sense and Numeration

Students quantify and measure objects, estimate mathematical quantities, and represent and communicate ideas in the language of mathematics.

Number Sense and Numeration

9. Concepts and Properties of Numbers – Students experience counting and measuring activities to develop intuitive sense about numbers, develop understanding about properties of numbers, understand the need for and existence of different sets of numbers, and investigate properties of special numbers.

- Develop an understanding of irrational, real and complex numbers.
- Use the (a+bi) and polar forms of complex numbers.
- Develop an understanding of the properties of the real and complex number systems and of the properties of special numbers including p, i, e, and conjugates.
- Apply their understanding of number systems to model and solve mathematical and applied problems.

10. Representation and Uses of Numbers - Students recognize that numbers are used in different ways such as counting, measuring, ordering and estimating, understand and produce multiple representations of a number, and translate among equivalent representations.

- Give decimal representations of rational and irrational numbers and coordinate
  and vector representations of complex numbers.
- Develop an understanding of more complex representations of numbers, including exponential and logarithmic expressions, and select an appropriate representation to facilitate problem solving.
- Determine when to use rational approximations and the exact values of numbers such as e, p and the irrationals.
- Apply estimation in increasingly complex situations.

- Select appropriate representations for numbers, including representation of rational and irrational numbers and coordinate or vector representation of complex numbers, in order to simplify and solve problems.

11. Number Relationships – Students investigate relationships such as equality, inequality, inverses, factors and multiples, and represent and compare very large and very small numbers.

- Compare and order real numbers and compare rational approximations to exact values.
- Express numerical comparisons as ratios and rates.
- Extend the relationships of primes, factors, multiples and divisibility in an algebraic setting.
- Express number relationships using positive and negative rational exponents, logarithms, and radicals.
- Apply their understanding of number relationships in solving problems.

Strand 4 Terms

Number Sense and Numeration

Real Number

Rational Number
Irrational Number
Complex Number

Integer

Natural Number

Equivalence Relation

Polar Coordinate

Counting

Measure

Order

Estimate

Equivalent

Exponential

Logarithm
Vector
Equality

Inequality

Inverse

Factor

Approximation

Ratio

Rate

Prime Number

Exponent

Lesson 4.1

1. A varies inversely with B. If A=12 when B=17, Find A when B=3.

a) 2.11
b) 4.25
c) 8.47
d) 68

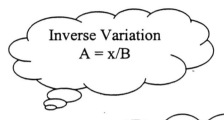

Inverse Variation
A = x/B

2. Order 3/7, ¼, ½, and 17/28 from greatest to least.

a) 1/2, 17/28, 3/7, ¼
b) 1/2, 3/7, ¼, 17/28
c) 17/28, ¾, ½, ¼
d) 17/28, ½, 3/7, ¼

The Quick Way – Put each value in your calculator and read it as though it were money. You can then easily put the numbers in order!

3. Which of the following is NOT an Integer?

a) 7
b) 1/2
c) 0
d) -3

4. The distance from the parking lot to the gymnasium is 769.24 feet. Express this distance using scientific notation.

a) 7.69×10^{-3}
b) 7.69×10^{-2}
c) 7.69×10^{2}
d) 7.69×10^{3}

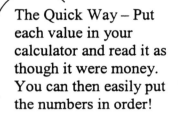

Press MODE, →, ENTER, CLEAR, (enter the number you want to convert), ENTER. After you get your answer, press MODE, ←, ENTER, CLEAR. This will return your calculator to its regular setting.

5. Danny's age is $\sqrt{171}$ years old. This is approximately how many years old?

a) 85.5
b) 17
c) 16.8
d) 13.1

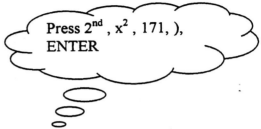

Press 2nd, x^2, 171,), ENTER

6. Mary's cookie recipe requires 3 parts sugar for 1 dozen cookies. If Mary makes 3 dozen cookies, what proportion could be used to represent the number of parts of sugar needed?

a) x/12 = 1/3
b) x/3 = 3/1
c) x/12 = ¼
d) x/3 = 1/4

7. Michelle's earnings vary directly with the number of hours she works. If she earns $102.00 for 8 hours of work, how much will she earn for 37 hours?

a) $446.25
b) $459.99
c) $471.75
d) $484.50

118

8.Six friends decided to sell baked goods to raise money. The friends, Amy, Beth, Cam, Dee, Eve, and Fred earned $684.26. Amy did ½ of the work, Beth ¼, Cam 1/8, Dee 1/16, Eve 1/32, and Fred 1/32. They agree to divide the money based upon the amount of work each person performed.

a) How much did each friend earn?

b) Express the distribution of funds utilizing a pie chart.

Lesson 4.2

1. Four out of nine students surveyed attended the last basketball game. What percent of the students surveyed attended the game?

a) 46%
b) 44%
c) 41%
d) 38%

2.

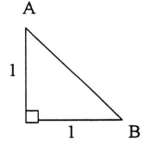

The distance from A to B is what kind of number?

a) Prime
b) Rational
c) Whole
d) Irrational

3. All reported answers on last Tuesdays Chemistry lab were to be rounded to the nearest hundredth. Darlene calculated a value of 462.8876. What is this value rounded to the nearest hundredth?

a) 500
b) 462.887
c) 462.888
d) 462.89

4. Which of the following is an irrational number?

a) 3/6
b) .6427
c) .642727272727...
d) .6427847697265...

5. Order the following from least to greatest

½, ¾, 5/8, 3/8

a) 3/8, ½, 5/8, ¾
b) 3/8, ¾, ½, 5/8
c) 1/2, 5/8, 3/8, ¾
d) 3/8, ½, ¾, 5/8

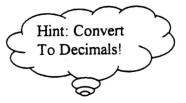

Hint: Convert To Decimals!

6. The Wilson's water bill varies directly with the number of gallons used. If their water bill is $37.70 for 290 gallons, how much would 560 gallons cost?

a) $4307.70
b) $75.40
c) $74.10
d) $72.80

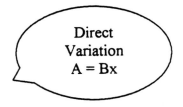

Direct Variation A = Bx

7. The pitch of a musical tone varies inversely with its wavelength. If a tone of 500 vibrations per second has a wavelength of 2 feet, find the wavelength of a tone with a pitch of 125 vibrations per second.

a) 125,000
b) 250
c) 62.5
d) 8

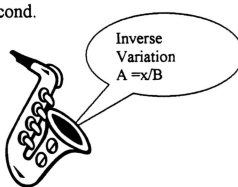

Inverse Variation A =x/B

8.Dave traveled to his grandmother's house for the Thanksgiving holiday. The highway distance between their homes is 250 miles. He traveled at a rate of 55 miles per hour.

a) How long did it take Dave to get to his grandmother's house?
b) During the ride home, Dave experienced post Thanksgiving traffic. If he traveled the same distance, but it took him 6 hours to get home, what was his average rate of speed? (Show all your work!)

Lesson 4.3

1. Express 28% as a fraction

a) 7/25
b) 14/25
c) 4/5
d) 6/7

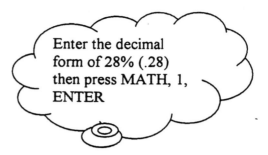

Enter the decimal form of 28% (.28) then press MATH, 1, ENTER

2. Which of the following is not equivalent to 1/7?

a) .1429
b) .17
c) 4/28
d) 14.29%

Convert each number (including the 1/7) into a decimal. There should be only 1 number not equivalent to 1/7.

3) Which of the following is NOT an Irrational number?
a) $\sqrt{2}$
b) π
c) .3689473
d) .2733434343434...

4. Which of the following is NOT a Rational number?

a) 5^2
b) -11
c) $\sqrt{2}$
d) 0

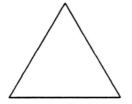

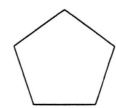

Strand 4: Number Sense and Numeration

5. In a recent survey, 8 out of 17 dentists prefer Brand X toothpaste. Which expression represents the percent of dentists who prefer Brand X?

a) 17/8 x 100
b) 9/17 x 100
c) 8/17 x 100
d) 17/9 x 100

To check, place every value into your TI-83

6. Marcus paid $9.00 for 15.5 gallons of gasoline. How much would 27 gallons cost?

a) $15.38
b) $15.68
c) $15.97
d) $46.50

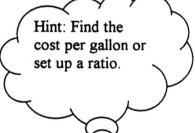

Hint: Find the cost per gallon or set up a ratio.

7. Order the following from least to greatest

   1/3, √3, .33, 3.3

a) 1/3, √3, .33, 3.3
b) √3, .33, 1/3, 3.3
c) .33, 1/3, √3, 3.3
d) 3.3, 1/3, √3, .33

Strand 4: Number Sense and Numeration

8.a) Shay has $334.00. If she deducts 20% for tithes and offerings, how much will she have left?

b) Of the money remaining, she gave $50 to Larry. What percent of the remaining amount did she give to Larry? Show all your work!

Strand 5:

Numerical and Algebraic Operations and Analytical Thinking

Students represent quantitative situations with numerical and algebraic symbolism and use analytic thinking to solve problems in significant contexts and applications.

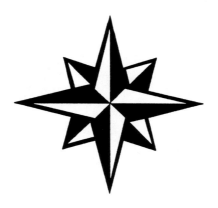

129

Numerical and Algebraic Operations and Analytic Thinking

12. Operations and their Properties – Students understand and use various types of operations (e.g., addition, subtraction, multiplication, division) to solve problems.

- Present and explain geometric and symbolic models for operations with real and complex numbers, and algebraic expressions.
- Compute with real numbers, complex numbers, algebraic expressions, matrices and vectors using technology and, for simple instances, with paper-and-pencil algorithms.
- Describe the properties of operations with numbers, algebraic expressions, vectors and matrices and make generalizations about the properties of given mathematical systems.
- Efficiently and accurately apply operations with real numbers, complex numbers, algebraic expressions, matrices, and vectors in solving problems.

13. Algebraic and Analytic Thinking – Students analyze problems to determine an appropriate process for solution, and use algebraic notations to model or represent problems.

- Identify important variables in a context, symbolize them, and express their relationships algebraically.
- Represent algebraic concepts and relationships with matrices, spreadsheets, diagrams, graphs, tables, physical models, vectors, equations and inequalities; and translate among the various representations.

Strand 5: Numerical and Algebraic Operations and Analytical Thinking

- Solve linear equations and inequalities algebraically and non-linear equations using graphing, symbol-manipulating or spreadsheet technology; and solve linear and non- linear systems using appropriate methods.
- Analyze problems that can be modeled by functions, determine strategies for solving the problems, and evaluate the adequacy of the solutions in the context of the problems.
- Explore problems that reflect the contemporary uses of mathematics in significant contexts and use the power of technology and algebraic and analytic reasoning to experience the ways mathematics is used in society.

Strand 5 Terms

Numerical and Algebraic Operations and Analytic Thinking

Numeric

Numerical Equation

Numerical Value

Geometric

Symbolic

Algebraic

Real Number

Rational Number

Irrational Number

Complex number

Integer

Natural Number

Equivalence Relation

Polar Coordinate

Matrix

Vector

Algorithm

Operation

Analytic

Variable

Spreadsheet

Diagram

Table

Chart

Graph

Model

Equation

Inequality

Linear

Nonlinear

Function

Reasoning

Lesson 5.1

1.Kevin is employed by Security Guard Safety Systems and must secure the perimeter of The Seaside Savings and Trust Bank. If the building is rectangular with the dimensions of 50 feet by 80 feet, what is its perimeter?

a) 30 feet
b) 130 feet
c) 260 feet
d) 400 feet

2.Mark took ¾ yards of red ribbon from a spool containing 7 ½ yards. How much ribbon was left on the spool?

a) 6 ¾ yards
b) 7 ¼ yards
c) 7 2/3 yards
d) 8 ¼ yards

3.At Candy's Candy Shop, super sour licorice costs $0.07 per cm. How much will Cornell pay for 87 mm?

a) $0.06
b) $0.61
c) $5.60
d) $6.09

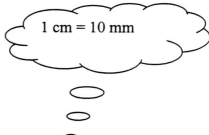

1 cm = 10 mm

4.Mary has 6 2/3 yards of denim. If she gives Karen 3 ½ yards, how much denim will she have left?

a) 3 3/5 yards
b) 3 1/6 yards
c) 3 yards
d) 2 ½ yards

5.Simone purchased 5 pounds of apples at $2.00 per pound, 4 pounds of oranges at $2.50 per pound and 6 pounds of pears at $1.00 per pound. How many pounds of fruit did Simone purchase?

a) 41 lbs.
b) 26 lbs.
c) 20 lbs.
d) 15 lbs.

6.The cost for a medium pizza is $7.00 plus $0.65 per topping. Diane's pizza cost $10.25. Which equation could be used to find "t", the number of toppings on Diane's medium pizza?

a) (7 + 0.65)t = $10.25
b) 7 + (0.65)t = $10.25
c) 7t + 0.65 = $7.00
d) 10.25t + 0.65 = $7.00

The price list at Tina's Yummy Yogurt Shop is given below.

Flavor	Price Per Scoop
1. Chocolate	$1.25
2. Strawberry	$1.00
3. Cherry	$0.75
4. Vanilla	$0.65
5. Peach	$0.50
6. Banana	$0.45

7. How much would a super sundae with one scoop of chocolate, one scoop of strawberry and two scoops of vanilla cost?

a) $3.55
b) $3.75
c) $4.00
d) $4.25

8. The Senior class raised $200 by selling chocolate and glazed donuts for the Prom. A total of 295 donuts were sold. If the chocolate donuts sold for $0.75 and the glazed donuts sold for $0.50, how many of each type of donut did they sell?

Systems of Equations!

Lesson 5.2

1.Andrea earns $6.25 per hour at the Icy Ice Cream Shop. She worked 7.25 hours on Tuesday, how much did she earn?

a) $45.31
b) $42.50
c) $13.50
d) $4.53

2.Ricardo spent 45 minutes on his Biology homework, 57 minutes on his English homework, 1 hour 17 minutes on his Geometry homework and 20 minutes on his History homework. What was the total time Ricardo spent on his homework?

a) 3hours, 59 minutes
b) 3hours, 49 minutes
c) 3hours, 19 minutes
d) 2hours, 59 minutes

3.Rochelle purchased 1 quart of milk to make molasses cookies. If the recipe requires 2 cups, how many cups of milk will Rochelle have left?

a) 2 cups
b) 4 cups
c) 8 cups
d) 16 cups

4. The surface area of a lateral cylinder is given by the formula

$$\textbf{Surface Area} = 2\pi rh + 2\,\pi r^2$$

If cylinder Y has a surface area of 444cm² and a radius of 8cm, what is its height?
(Use $\pi = 3.14$)

a) 8.25cm
b) 2.52cm
c) 2cm
d) 0.84cm

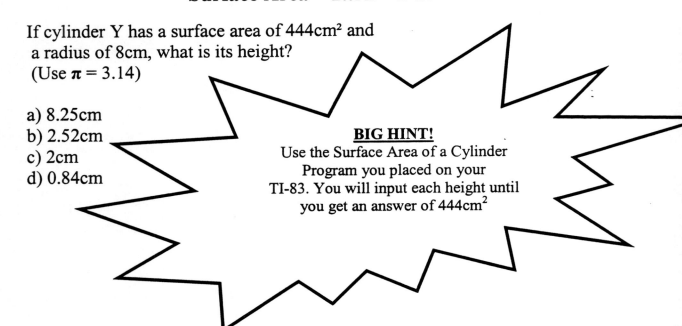

BIG HINT!
Use the Surface Area of a Cylinder Program you placed on your TI-83. You will input each height until you get an answer of 444cm²

5. Pete's Plastic Shop ordered a shipment of cubical containers. If a container has a side length of 7 inches, what is its volume?

a) 343 cm³
b) 294 cm³
c) 98 cm³
d) 16 cm³

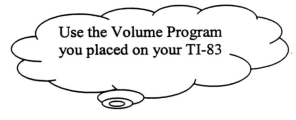

Use the Volume Program you placed on your TI-83

6. Approximately how many ¾ inch strips can be made from a piece of ribbon 16 2/3 inches long?

a) 12
b) 15
c) 17
d) 22

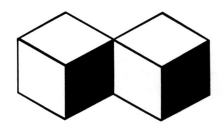

7. The following fruit salad recipe serves 6 people

Recipe
5 Apples
5 Oranges
2 Pineapples
3 Cups Cherries

Combine ingredients and mix well

Think in terms of ratios, if 2 Pineapples serves 6 people, then "x" pineapples serves 20 people.

How many pineapples would be needed to feed 20 people?

a) 2 5/6
b) 4 ½
c) 6 2/3
d) 10

8. Alonzo has $21.40 in dimes and quarters. If he has 100 coins how many dimes and quarters does he have? Show all work.

Systems of Equations!

141

Lesson 5.3

1. It takes Jane 12 ounces of paint, to paint one door. If she purchases two gallons of paint, approximately how many doors will she be able to paint?

a) 6 doors
b) 11 doors
c) 14 doors
d) 21 doors

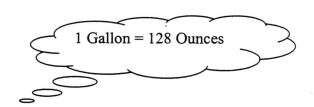

1 Gallon = 128 Ounces

2. David walked 6 yards from point A to Point B, 5 yards from Point B to Point C and 3 yards from Point C to Point D. How many feet did David walk?

a) 178 feet
b) 114 feet
c) 42 feet
d) 14 feet

3. The volume of a rectangular prism is given by the formula
Volume = length x width x height = lwh
What is the volume of a rectangular prism 5cm long, 6cm wide and 7cm high?

a) 18 cm³
b) 65 cm³
c) 72 cm³
d) 210 cm³

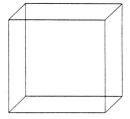

4. A 64ounce bottle of cola costs $1.92 and a bag of chips cost $0.59. What is the cost of the cola per ounce?

a) $2.51
b) $1.37
c) $0.62
d) $0.03

5. Anita needs to save $250.00 for a spring break trip. She already has $35 in her piggy bank. If she saves $15.00 per week, which expression could be used to represent x, the number of weeks it would take her to save at least $250.00?

a) $35 + 15X \geq 250$
b) $(35 + 15)X \geq 250$
c) $35X + 15 \geq 250$
d) $35 + 15X \leq 250$

6. Lisa earned $82.49 last week. With her earnings, she spent $17.24 on a sweatshirt and $25.62 on a pair of shoes. How much money does Lisa have left?

a) $90.87
b) $74.11
c) $65.25
d) $39.63

7. Ricky must calculate the volume of the cylinder that he will be using in his Chemistry experiment. If the cylinder has a height of 14 cm and a radius of 4 cm, what is its volume?

a) 784 cm³
b) 704 cm³
c) 448 cm³
d) 224 cm³

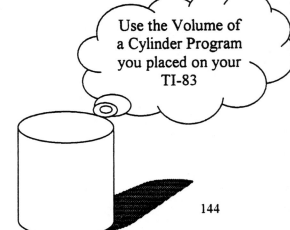

Use the Volume of a Cylinder Program you placed on your TI-83

144

8.

Kelly's Room

In Kelly's rectangular room, the length is 4 more than twice the width. The perimeter is 56cm. Find the length and the width of Kelly's room. Show all supporting calculations.

Systems of Equations!

Strand 6:

Probability and Discrete Mathematics

Students deal with uncertainty, make informed decisions based on evidence and expectations, exercise critical judgment about conclusions drawn from data, and apply mathematical models to real-world phenomena.

Strand 6: Probability and Discrete Mathematics

Strand 6: Probability and Discrete Mathematics

Probability and Discrete Mathematics

14. Probability – Students develop an understanding of the notion of certainty and of probability as a measure of degree of likelihood that can be assigned to a given event based on the knowledge available, and make critical judgments about claims that are made in probabilistic situations.

- ξ Develop an understanding of randomness and chance variation and describe chance and certainty in the language of probability.
- ξ Give a mathematical definition of probability and determine the probabilities of more complex events, and generate and interpret probability distributions.
- ξ Analyze events to determine their dependence or independence and calculate probabilities of compound events.
- ξ Use sampling and simulations to determine empirical probabilities and, when appropriate, compare them to the corresponding theoretical probabilities; understand and apply the law of large numbers.
- ξ Conduct probability experiments and simulations to model and solve problems, including compound events.

15. Discrete Mathematics – Students investigate practical situations such as scheduling, routing, sequencing, networking, organizing and classifying, and analyze ideas like recurrence relations, induction, iteration, and algorithm design.

- ξ Derive and use formulas for calculating permutations and combinations.
- ξ Use sets and set relationships to represent algebraic and geometric concepts.

Strand 6: Probability and Discrete Mathematics

ξ Use vertex-edge graphs to solve network problems such as finding circuits, critical paths, minimum spanning trees, and adjacency matrices.

ξ Analyze and use discrete ideas such as induction, iteration and recurrence relations.

ξ Describe and analyze efficient algorithms to accomplish a task or solve a problem in a variety of contexts including practical, mathematical and computer-related situations.

ξ Use discrete mathematics concepts as described above to model situations and solve problems; and look for whether or not there is a solution (existence problems), determine how many solutions there are (counting problems), and decide upon a best solution (optimization problems).

Strand 6: Probability and Discrete Mathematics

Strand 6 Terms

Probability and Discrete Mathematics

Probability

Certainty

Event

Random
Variation

Event

Probabilistic

Dependence

Independence

Simulation

Empirical
Theoretical

 Model

Discrete

Scheduling

Sequence

Network
Recurrence

Induction

Iteration

Algorithm
Permutation

Combination

Set

Vertex edge graph
Critical paths
Minimum spanning trees
Adjacency matrix
Existence problems
Optimization problems

Strand 6: Probability and Discrete Mathematics

Strand 6: Probability and Discrete Mathematics

Lesson 6.1

1.George has 3 nickels, 2 quarters, 2 dimes, and 5 pennies in his pocket. If Tom selects one coin randomly, what is the probability that it will be a nickel?

a) 1/6
b) 1/4
c) 1/3
d) 1/2

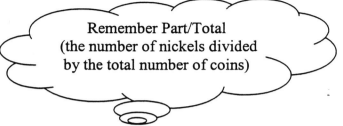
Remember Part/Total
(the number of nickels divided by the total number of coins)

2.Jacqueline is attempting to coordinate her hats and shoes. If she has six hats and four pairs of shoes, how many different combinations can she create?

a) 12
b) 14
c) 18
d) 24

3.If two coins are flipped at the same time, what is the probability that one heads and one tails will appear?

a) 1
b) 1/2
c) 1/4
d) 1/6

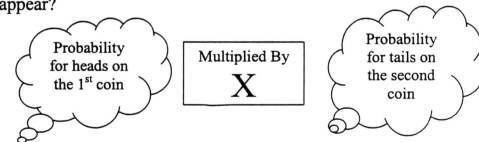
Probability for heads on the 1st coin

Multiplied By
X

Probability for tails on the second coin

4.If a coin and a die are tossed, what is the probability that a number less than 5 and a heads would appear?

a) 1/3
b) 1/4
c) 1/6
d) 1/8

Strand 6: Probability and Discrete Mathematics

5. In Mrs. Palmers' class, there are 14 sophomores, 18 juniors, and 18 seniors. What is the probability that a student chosen at random will **NOT** be a senior?

a) 3/5
b) 18/50
c) 16/25
d) 1/36
e) 1/64

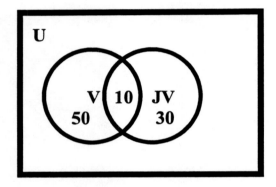

U = Set of 300 students at Murray High School
V = Varsity Players
JV = Junior Varsity Players

How many students play neither varsity nor junior varsity?

a) 210 students
b) 200 students
c) 100 students
d) 90 students

Strand 6: Probability and Discrete Mathematics

7. The Roberts family purchased a home four years ago for $75,000. The value of the home has appreciated each year by 15%. How much is the home worth today?

a) $131,175.47
b) $120,000.00
c) $114,065.63
d) $86,250.00

Avoid the mistake of rushing!!! Add 15% to the price of the home each year. NOTE, each year you will be adding the NEW 15% to the APPRECIATED value of the home.

8. The following candidates are running for President, Vice President and Treasurer of the Student Council:

President: Matthew, Mark, David, and Esther
Vice - President: Luke, John, and Ruth
Treasurer: Rachael, Leah, and Naomi
(Males- Matthew, Mark, David, Luke, and John)
(Females- Naomi, Rachael, Leah, Esther, and Naomi)

a) List all the possible outcomes
b) What is the probability that a female will be elected to all three positions?
Show all supporting calculations

Strand 6: Probability and Discrete Mathematics

Lesson 6.2

1. How many outfits (1 top/1 bottom) can Mackenzie create from 6 shirts and 6 pairs of slacks?

a) 12
b) 25
c) 30
d) 36

2. Roy is attempting to determine the value of his luxury automobile. According to a leading auto magazine a luxury car like Roy's will depreciate in value 10% each year. When the car was new, its value was $30,000. If the car is three years old, what is its value today?

a) $27,000.00
b) $22,576.35
c) $21,870.00
d) $3,000.00

3. Jene' has a bag of marbles. There are 15 red, 30 green, 7 blue, 8 gray, 6 yellow, and 4 clear marbles. If Jene' reaches into her bag without looking what is the probability that she will draw a green marble?

a) 3/35
b) 1/10
c) 3/7
d) 4/35

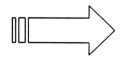

4.A locker contains 4 math books, 3 English books, and 5 history books. If one book is randomly selected, what is the probability that it will NOT be a math book?

a) 2/3
b) 1/3
c) 1/6
d) 2/18

5.If two dice are tossed, what is the probability that a 1 will appear on both dice?

a) 36
b) 12
c) 1/3
d) 1/36

6.What is the probability that a card chosen randomly from a standard deck of 52 will not be a face card (Jack , Queen, King)?

a) 3/52
b) 3/13
c) 4/13
d) 10/13

7.If two coins are tossed, what is the probability that heads would appear on both coins?

a) 2
b) 1/2
c) 1/4
d) 1/8

Team Rosters

8. Varsity Golf =A = {Victor, Ezekiel, Paul, David, Karen, Saul, Tom, Tim, Ralph, Walter and Harold}

Cross Country =B= {Ezekiel, Paul, David, Karen, Saul, Tom and Sharome}

a) Find A∩B. Explain what this set represents.

b) Find A   B. Explain what this set represents.

"∩" (intersection) asks the question "which elements do these sets have in common?"

"U"(union) says "bring all the elements together (and NO, you do not list anything/anyone twice!)

Strand 6: Probability and Discrete Mathematics

Strand 6: Probability and Discrete Mathematics

Lesson 6.3

1. How many three digit numbers can be formed using the digits 1, 2, & 3 (A number can be repeated)

a) 27
b) 9
c) 6
d) 4

2. Jean has decided to open her new business "Just Desserts". To minimize supply cost, she will only offer 5 icings (chocolate, vanilla, strawberry, peanut butter and coconut) and 5 different cake batter flavors (chocolate, vanilla, strawberry, butter and carrot). How many different cakes can Jean create?

a) 36
b) 30
c) 25
d) 11

3. How many meals can be created from 3 meats, 5 vegetables, and 2 desserts? (A meal consists if 1 meat, 1 vegetable and 1 dessert)

a) 12
b) 20
c) 30
d) 60

4.There are 35 students on the Student Council. Of the students, 3 are in the 9th grade, 7 are in the 10th grade, 11 are in the 11th grade and the remaining students are 12th graders. What is the probability that a student chosen at random will be a 11th or 12th grader?

a) 7/8
b) 5/7
c) 1/2
d) 11/35

5.A nickel, dime and quarter are flipped at the same time. What is the probability that all three coins will show heads?

a) 3
b) 2/3
c) 1/3
d) 1/8

6.What is the probability of rolling two dice and getting a number less than three on the first and a number greater than two on the second?

a) 6
b) 8/6
c) 1
d) 2/9

7.If a die and a coin are tossed, what is the probability that a 4 and a tails will appear?

a) 12
b) 1/3
c) 1/4
d) 1/12

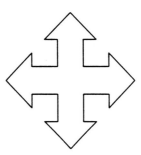

8. Alice, Beth, Carmen, and Dana will serve as bridesmaids in Rebecca and Paul's wedding. Daniel, Shadrach, Meschach, and Abednego will serve as groomsmen. In the wedding procession one male and one female (bridesmaid/groomsman) is to walk down the isle.

a) List all possible combinations of who could walk down the isle together.

b) What is the probability that the two individuals walking down the isle together will have the same first letter in their name?

Glossary and Answer Keys

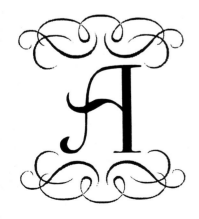

Glossary
Strand 1 Terms

1. **Sequence** - A following of one thing after another; succession. An order of succession; an arrangement.

2. **Series** - A number of objects or events arranged or coming one after the other in succession.

3. **Recursion** - An expression whereby each term is determined by application of a formula to preceding terms. A formula that generates the successive terms of a recursion.

4. **Matrix** - A rectangular array of numeric or algebraic quantities subject to mathematical operations. Something resembling such an array, as in the regular formation of elements into columns and rows.

5. **Graph** - A diagram that exhibits a relationship, often functional, between two sets of numbers as a set of points having coordinates determined by the relationship. Also called **plot**. A pictorial device, such as a pie chart or bar graph, used to illustrate quantitative relationships. Also called **chart**.

6. **Vector** - A quantity, such as velocity, completely specified by a magnitude and a direction. A one-dimensional array.

7. **Inference** - The act or process of deriving logical conclusions from premises known or assumed to be true. The act of reasoning from factual knowledge or evidence.

8. **Prediction** - Something foretold or predicted; a prophecy.

9. **Graphic Pattern** – A visual representation where succeeding terms follow a prescribed sequence.

10. **Numeric Pattern** - A numeral representation where succeeding terms follow a prescribed sequence.

11. **Linear Pattern** - A representation when graphed, represents a line.

12. **Quadratic Pattern** - A representation when graphed, represents a parabola.

13. **Family of Functions** – Graphs and equations that have at least one characteristic in common.

14. **Rate** - A quantity measured with respect to another measured quantity: the plumber is paid at *a rate of speed of 60 dollars per hour*. A measure of a part with respect to a whole; a proportion: *the mortality rate; a tax rate.*

15. **Continuity** - An uninterrupted succession or flow; a coherent whole.

16. **Limit** - The point, edge, or line beyond which something cannot or may not proceed. The boundary surrounding a specific area; bounds: *within the city limits.* A confining or restricting object, agent, or influence. The greatest or least amount, number, or extent allowed or possible: *a withdrawal limit of $200; no minimum age limit.* A number or point L that is approached by a function $f(x)$ as x approaches a if, for every positive number ε, there exists a number δ such that $|f(x)-L| < \varepsilon$ if $0 < |x-a| < \delta$. Also called **limit point, point of accumulation**.

17. **Distribution** - The act of distributing or the condition of being distributed; apportionment. Something distributed; an allotment. The act of dispersing or the condition of being dispersed; diffusion. Division into categories; classification. A set of numbers and their frequency of occurrence collected from measurements over a statistical population. A generalized function used in the study of partial differential equations.

18. **Deviation** - The act of deviating or turning aside. The difference, especially the absolute difference, between a number in a set and the mean of the set.

19. **Function** - A variable so related to another that for each value assumed by one there is a value determined for the other. A rule of correspondence between two sets such that there is a unique element in the second set assigned to each element in the first set.

20. **Linear** – Of, relating to, or resembling a line; straight. In, of, describing, described by, or related to a straight line. Having only one dimension.

21. **Linear equation** - An equation of the first degree between two variables; so called because every such equation may be considered as representing a right line.

22. **Quadratic** - Of, relating to, or containing quantities of the second degree.

23. **Exponential** - Of or relating to an exponent. Containing, involving, or expressed as an exponent. Expressed in terms of a designated power of e, the base of natural logarithms.

24. Trigonometric Function - A function of an angle expressed as the ratio of two of the sides of a right triangle that contains that angle; the sine, cosine, tangent, cotangent, secant, and cosecant. Also called **circular function**.

25. Composition of functions – If f and g are functions such that the range of g is a subset of the domain of f. Then the composite function fog is given by [fog](x) = f[g(x)].

26. Piece wise functions – A function defined by several functions.

27. Recursively defined functions - A function whereby each term is determined by application of a formula to preceding terms. A function that generates the successive terms of a recursion.

28. Inverse functions – Two functions whose composition is the identity function.

Strand 2 Terms

1. Loci - The set or configuration of all points whose coordinates satisfy a single equation or one or more algebraic conditions. The line traced by a point which varies its position according to some determinate law; the surface described by a point or line that moves according to a given law.

2. Graph - A diagram that exhibits a relationship, often functional, between two sets of numbers as a set of points having coordinates determined by the relationship. Also called **plot**. A pictorial device, such as a pie chart or bar graph, used to illustrate quantitative relationships. Also called **chart**.

3. Distribution - The act of dispersing or the condition of being dispersed; diffusion. Division into categories; classification. A set of numbers and their frequency of occurrence collected from measurements over a statistical population.

4. Transformation - Replacement of the variables in an algebraic expression by their values in terms of another set of variables. A mapping of one space onto another or onto itself.

5. Coordinate - Any of a set of two or more numbers used to determine the position of a point, line, curve, or plane in a space of a given dimension with respect to a system of lines or other fixed references.

6. Synthetic – Something determined not true by observation not solely by its components.

7. Dimension - A measure of spatial extent, especially width, height, or length. Extent or magnitude; scope. A physical property, such as mass, length, time, or a combination thereof, regarded as a fundamental measure or as one of a set of fundamental measures of a physical quantity: Velocity has the dimensions of length and time.

8. Isometry -Equality of measure. A function between metric spaces which preserves distances, such as a rotation or translation in a plane.

9. Coordinate - Any of a set of two or more numbers used to determine the position of a point, line, curve, or plane in a space of a given dimension with respect to a system of lines or other fixed references.

10. Congruence - Agreement, harmony, conformity, or correspondence.

11. Similarity - The quality or condition of being alike; resemblance. A corresponding aspect or feature; equivalence.

12. Perpendicular - Intersecting at or forming right (90 degree) angles.

13. Incidence - The direction in which a body, or a ray of light or heat, falls on any surface. <u>Angle of incidence</u> - angle which a ray of light, or the line of incidence of a body, falling on any surface, makes with a perpendicular to that surface; also formerly, the complement of this angle.

14. Polar Coordinates - coordinates made up of a radius vector and its angle of inclination to another line, or a line and plane.

15. Vector - A quantity, such as velocity, completely specified by a magnitude and a direction. A one-dimensional array.

16. Limit - The point, edge, or line beyond which something cannot or may not proceed. The boundary surrounding a specific area. A confining or restricting object, agent, or influence. A number or point L that is approached by a function $f(x)$ as x approaches a if, for every positive number ε, there exists a number δ such that $|f(x)-L| < \varepsilon$ if $0 < |x-a| < \delta$. Also called **limit point, point of accumulation**.

18. maximum –The greatest possible quantity or degree. The greatest quantity or degree reached or recorded; the upper limit of variation. The greatest value assumed by a function over a given interval. The largest number in a set.

19. minimum - The least possible quantity or degree. The lowest degree or amount reached or recorded; the lower limit of variation. The smallest number in a finite set of numbers. A value of a function that is less than any other value of the function over a specific interval.

20. Inflection - A turning or bending away from a course or position of alignment. Alteration in pitch or tone.

21. Symmetry - Exact correspondence of form and constituent configuration on opposite sides of a dividing line or plane or about a center or an axis.

22. Measurement - The act of measuring or the process of being measured. A system of measuring: *measurement in miles*. The dimension, quantity, or capacity determined by measuring: *the measurements of a room*.

23. Precision - The state or quality of being precise; exactness. The ability of a measurement to be consistently reproduced. The number of significant digits to which a value has been reliably measured.

24. Metric - A system of weights and measures originating in France, the use of which is required by law in many countries, and permitted in many others, including the United States and England. The principal unit is the meter. From this are formed the are, the liter, the stere, the gram, etc. These units, and others derived from them, are divided decimally, and larger units are formed from multiples by 10, 100, 1,000, and 10,000. The successive multiplies are designated by the prefixes, deca-, hecto-, kilo-, and myria-; successive parts by deci-, centi-, and milli-. The prefixes mega- and micro- are sometimes used to denote a multiple by one million, and the millionth part, respectively. See the words formed with these prefixes in the Vocabulary. For metric tables, see p. 1682.

25. Angle - The figure formed by two lines diverging from a common point. The figure formed by two planes diverging from a common line. The rotation required to superimpose either of two such lines or planes on the other. The space between such lines or surfaces.

26. Degree - A planar unit of angular measure equal in magnitude to $1/360$ of a complete revolution. The greatest sum of the exponents of the variables in a term of a polynomial or polynomial equation. A unit division of a temperature scale.

27. Radian - A unit of angular measurement.

28. Parallel - Being an equal distance apart everywhere. Of, relating to, or designating two or more straight coplanar lines that do not intersect. Of, relating to, or designating two or more planes that do not intersect. Of, relating to, or designating a line and a plane that do not intersect. Of, relating to, or designating curves or surfaces everywhere equidistant.

29. Length - The state, quality, or fact of being long.

30. Mass - A unified body of matter with no specific shape. The physical volume or bulk of a solid body. A property of matter equal to the measure of an object's resistance to changes in either the speed or direction of its motion.

31. Temperature – The degree of hotness or coldness of a body or environment. A measure of the average kinetic energy of the particles in a sample of matter, expressed in terms of units or degrees designated on a standard scale.

32. Area - The extent of a 2-dimensional surface enclosed within a boundary.

33. Volume - Dimensions; compass; space occupied, as measured by cubic units, that is, cubic inches, feet, yards, etc.; mass; bulk.

34. Precision - The state or quality of being precise; exactness. The ability of a measurement to be consistently reproduced. The number of significant digits to which a value has been reliably measured.

35. Reasoning - Evidence or arguments used in thinking or argumentation. Use of reason, especially to form conclusions, inferences, or judgments.

36.Indirect Measurements - A course of reasoning showing that a certain result is a necessary consequence of assumed premises; -- these premises being definitions, axioms, and previously established propositions.

37.Sine - In a right triangle, the ratio of the length of the side opposite an acute angle to the length of the hypotenuse.

38.Cosine - In a right triangle, the ratio of the length of the side adjacent to an acute angle to the length of the hypotenuse.

39.Tangent - Making contact at a single point or along a line; touching but not intersecting. The trigonometric function of an acute angle in a right triangle that is the ratio of the length of the side opposite the angle to the length of the side adjacent to the angle.

*Include all measures listed on the reference sheet within the Strand 2 terms.

Strand 3 Terms

1. Data - Factual information, especially information organized for analysis or used to reason or make decisions.

2. Survey - To examine or look at comprehensively.

3. Sample - A portion, piece, or segment that is representative of a whole. An entity that is representative of a class; a specimen. A set of elements drawn from and analyzed to estimate the characteristics of a population. Also called **sampling**.

4. Simulation - Imitation or representation, as of a potential situation or in experimental testing. Representation of the operation or features of one process or system through the use of another.

5. Table - A set of data arranged in rows and columns.

6. Chart - An outline map on which specific information, such as scientific data, can be plotted. A sheet presenting information in the form of graphs or tables.

7. Graph - A diagram that exhibits a relationship, often functional, between two sets of numbers as a set of points having coordinates determined by the relationship. Also called **plot**. A pictorial device, such as a pie chart or bar graph, used to illustrate quantitative relationships. Also called chart.

8. Spreadsheet - A piece of paper with rows and columns for recording financial data for use in comparative analysis.

9. Database - A collection of data arranged for ease and speed of search and retrieval. Also called **data bank**.

10. Skew - To take an oblique course or direction. To look obliquely or sideways. To turn or place at an angle. To give a bias to; distort.

11. Bias - A statistical sampling or testing error caused by systematically favoring some outcomes over others.

12. Distribution - A set of numbers and their frequency of occurrence collected from measurements over a statistical population. The act of distributing or the condition of being distributed; apportionment. Something distributed; an allotment. The act of dispersing or the condition of being dispersed; diffusion. Division into categories; classification.

13. Mean - The average value of a set of numbers. A number that typifies a set of numbers, such as a geometric mean or an arithmetic mean.

14. Median - Relating to, located in, or extending toward the middle. Relating to or constituting the middle value in a distribution. The middle value in a distribution, above and below which lie an equal number of values. A line that

joins a vertex of a triangle to the midpoint of the opposite side. The line that joins the midpoints of the nonparallel sides of a trapezoid.

15. Mode - The value or item occurring most frequently in a series of observations or statistical data. The number or range of numbers in a set that occurs the most frequently.

16. Variability - The quality, state, or degree of being variable or changeable.

17. Correlation - The simultaneous change in value of two numerically valued random variables: *the positive correlation between years of education and yearly income; the negative correlation between age and normal vision.*

18. Inference - The act or process of deriving logical conclusions from premises known or assumed to be true. The act of reasoning from factual knowledge or evidence.

19. Prediction - Something foretold or predicted; a prophecy.

20. Hypothesis - A tentative explanation for an observation, phenomenon, or scientific problem that can be tested by further investigation. Something taken to be true for the purpose of argument or investigation; an assumption. The antecedent of a conditional statement.

21. Modeling - To make or construct a model of. To plan, construct, or fashion according to a model. To make conform to a chosen standard: To make by shaping a plastic substance: *model clay.* In painting, drawing, and photography, to give a three-dimensional appearance to by shading or highlighting.

21. **Confidence Interval** - A statistical range with a specified probability that a given parameter lies within the range.

22. Interpolate - To estimate a value of (a function or series) between two known values. To make insertions or additions.To insert or introduce between other elements or parts. To insert (material) into a text. To change or falsify (a text) by introducing new or incorrect material.

23. Extrapolate - To infer or estimate by extending or projecting known information. To estimate (a value of a variable outside a known range) from values within a known range by assuming that the estimated value follows logically from the known values.

24. Interval – A set of numbers containing all the numbers between two given numbers.

25. Nominal – Information given in the form of names.

26. Ordinal – Expresses order or succession. A number used to indicate order ex. 1st, 5th or 38th.

27. Ratio – A relation between two similar things. The quotient of one quantity divided by another.

Strand 4 Terms

1. Real Number - One of the infinitely divisible range of values between positive and negative infinity, used to represent continuous physical quantities such as distance, time and temperature.

Between any two real numbers there are infinitely many more real numbers. The integers ("counting numbers") are real numbers with no fractional part and real numbers ("measuring numbers") are complex numbers with no imaginary part. Real numbers can be divided into rational and irrational numbers.

2. Rational Number - A number capable of being expressed as an integer or a quotient of integers, excluding zero as a denominator.

3. Irrational Number - Any real number that cannot be expressed as a ratio between two integers.

4. Complex Number - Any number of the form $a + bi$, where a and b are real numbers and i is an imaginary number whose square equals -1.

5. Integer - A member of the set of positive whole numbers $\{1, 2, 3,... \}$, negative whole numbers $\{-1, -2, -3,... \}$, and zero $\{0\}$.

6. Natural Number - One of the set of positive whole numbers; a positive integer.

7. Equivalence Relation - A reflexive, symmetric, and transitive relationship between elements of a set, such as congruence for the set of all triangles in a plane.

8. Polar Coordinate - Either of two numbers that locate a point in a plane by its distance from a fixed point on a line and the angle this line makes with a fixed line.

9. Counting - To name or list (the units of a group or collection) one by one in order to determine a total; number.

10. Measure - Dimensions, quantity, or capacity as ascertained by comparison with a standard. A reference standard or sample used for the quantitative comparison of properties: *The standard kilogram is maintained as a measure of mass.* A unit specified by a scale, such as an inch, or by variable conditions, such as a day's march. A system of measurement, such as the metric system.

11. Order - To reduce to a methodical arrangement; to arrange in a series, or with reference to an end.

12. Estimate - To calculate approximately (the amount, extent, magnitude, position, or value of something. The act of evaluating or appraising. A tentative evaluation or rough calculation, as of worth, quantity, or size. A statement of the approximate cost of work to be done, such as a building project or car repairs. A judgment based on one's impressions; an opinion.

13. Equivalent - Equal, as in value, force, or meaning. Having similar or identical effects. Being essentially equal, all things considered. Capable of being put into a one-to-one relationship. Used of two sets. Having virtually identical or corresponding parts. Of or relating to corresponding elements under an equivalence relation..

14. Exponential - Containing, involving, or expressed as an exponent. Expressed in terms of a designated power of e, the base of natural logarithms.

15. Logarithm - The power to which a base, such as 10, must be raised to produce a given number. If $n^x = a$, the logarithm of a, with n as the base, is x; symbolically, $\log_n a = x$. For example, $10^3 = 1,000$; therefore, $\log_{10} 1,000 = 3$. The kinds most often used are the common logarithm (base 10), the natural logarithm (base e), and the binary logarithm (base 2).

16. Vector - A quantity, such as velocity, completely specified by a magnitude and a direction. A one-dimensional array.

17. Equality - The state or quality of being equal. A statement, usually an equation, that one thing equals another.

18. Inequality - The condition of being unequal. An instance of being unequal. Lack of equality, as of opportunity, treatment, or status. An algebraic relation showing that a quantity is greater than or less than another quantity

19. Inverse - One of a pair of elements in a set whose result under the operation of the set is the identity element, especially: The reciprocal of a designated quantity. Also called **multiplicative inverse**. The negative of a designated quantity. Also called **additive inverse**. Reversed in order, nature, or effect. Of or relating to an inverse or an inverse function. Turned upside down; inverted. Something that is opposite, as in sequence or character; the reverse.

20. Factor - One of two or more quantities that divides a given quantity without a remainder. For example, 2 and 3 are factors of 6; a and b are factors of ab. A quantity by which a stated quantity is multiplied or divided, so as to indicate an increase or decrease in a measurement: *The rate increased by a factor of ten.*

21. Approximation - An inexact result adequate for a given purpose.

22. Ratio - Relation in degree or number between two similar things. The relation between two quantities expressed as the quotient of one divided by the other: *The ratio of 7 to 4 is written 7:4 or 7/4.*

23. Rate - A quantity measured with respect to another measured quantity: *a rate of speed of 60 miles an hour.* A measure of a part with respect to a whole. The cost per unit of a commodity or service: *postal rates.* A charge or payment calculated in relation to a particular sum or quantity: *interest rates.*

24. Prime Number - A positive integer not divisible without a remainder by any positive integer other than itself and one.

25. Exponent - A number, letter, or any quantity written on the right hand of and above another quantity, and denoting how many times the latter is repeated as a factor to produce the power indicated; Note: thus a^2 denotes the second power, and a^x, the xth power, of a (2 and x being the exponents). A fractional exponent, or index, is used to denote the root of a quantity. Thus, $a^{1/3}$ denotes the third or cube root of a. A number that when multiplied by itself an indicated number of times forms a product equal to a specified number. For example, a fourth root of 4 is $\sqrt{}$ 2. Also called **nth root**. A number that reduces a polynomial equation in one variable to an identity when it is substituted for the variable. A number at which a polynomial has the value zero.

Strand 5 Terms

1. Numeric - Belonging to number; denoting number; consisting in numbers; expressed by numbers, and not letters; as, numerical characters; a numerical equation; a numerical statement. Note: Numerical, as opposed to algebraic, is used to denote a value irrespective of its sign; thus, -5 is numerically greater than -3, though algebraically less.

2. Numerical Equation - An equation which has all quantities except the known expressed in numbers.

3. Numerical Value - An equation or expression that is deduced by substituting numbers for the letters, and reducing.

4. Geometric - Of or relating to geometry and its methods and principles. Increasing or decreasing in a geometric progression. Using simple geometric forms such as circles and squares in design and decoration. Of or relating to properties in algebraic geometry involving algebraically closed fields.

5. Symbolic - Of, relating to, or expressed by means of symbols or a symbol.

6. Algebraic - Designating an expression, equation, or function in which only numbers, letters, and arithmetic operations are contained or used. Indicating or restricted to a finite number of operations involving algebra.

7. Real Number - One of the infinitely divisible range of values between positive and negative infinity, used to represent continuous physical quantities such as distance, time and temperature. Between any two real numbers there are infinitely many more real numbers. The integers ("counting numbers") are real numbers with no fractional part and real numbers ("measuring numbers") are complex numbers with no imaginary part. Real numbers can be divided into rational and irrational numbers.

8. Rational Number - A number capable of being expressed as an integer or a quotient of integers, excluding zero as a denominator.

9. Irrational Number - Any real number that cannot be expressed as a ratio between two integers

10. Complex number - Any number of the form $a + bi$, where a and b are real numbers and i is an imaginary number whose square equals -1.

11.Integer - A member of the set of positive whole numbers {1, 2, 3,... }, negative whole numbers {-1, -2, -3,... }, and zero {0}.

12.Natural Number - One of the set of positive whole numbers; a positive integer.

13. Equivalence Relation - A reflexive, symmetric, and transitive relationship between elements of a set, such as congruence for the set of all triangles in a plane.

14. Polar Coordinate - Either of two numbers that locate a point in a plane by its distance from a fixed point on a line and the angle this line makes with a fixed line.

15. Matrix - A rectangular array of numeric or algebraic quantities subject to mathematical operations. Something resembling such an array, as in the regular formation of elements into columns and rows.

16. Vector - A quantity, such as velocity, completely specified by a magnitude and a direction. A one-dimensional array. An element of a vector space.

17. Algorithm - A step-by-step problem-solving procedure, especially an established, recursive computational procedure for solving a problem in a finite number of steps.

18. Operation - A process or action, such as addition, substitution, transposition, or differentiation, performed in a specified sequence and in accordance with specific rules.

19. Analytic - Using, subjected to, or capable of being subjected to a methodology involving algebra or other methods of mathematical analysis. Proving a known truth by reasoning from that which is to be proved.

20. Variable - Likely to change or vary; subject to variation; changeable. Having no fixed quantitative value. A quantity capable of assuming any of a set of values. A symbol representing such a quantity. For example, in the expression $a^2 + b^2 = c^2$, a, b, and c are variables.

21. Spreadsheet - A piece of paper with rows and columns for recording financial data for use in comparative analysis. An accounting or bookkeeping program that displays data in rows and columns on a screen.

22. Diagram - A plan, sketch, drawing, or outline designed to demonstrate or explain how something works or to clarify the relationship between the parts of a whole. A graphic representation of an algebraic or geometric relationship. A chart or graph.

23. Table - A set of data arranged in rows and columns.

24. Chart - An outline map on which specific information, such as scientific data, can be plotted. A sheet presenting information in the form of graphs or tables.

25. Graph - A diagram that exhibits a relationship, often functional, between two sets of numbers as a set of points having coordinates determined by the relationship. Also called **plot**. A pictorial device, such as a pie chart or bar graph, used to illustrate quantitative relationships. Also called **chart**.

26. Model – To make a copy or a pattern; to design or imitate forms; as, to model in wax.

27. Equation - A making equal; equal division; equality; equilibrium. An expression of the condition of equality between two algebraic quantities or sets of quantities, the sign = being placed between them; as, a binomial equation; a quadratic equation; an algebraic equation; a transcendental equation; an exponential equation; a logarithmic equation; a differential equation, etc.

28. Inequality - The condition of being unequal. An instance of being unequal. Lack of equality, as of opportunity, treatment, or status. An algebraic relation showing that a quantity is greater than or less than another quantity.

29. Linear - Of, relating to, or resembling a line; straight. In, of, describing, described by, or related to a straight line. Having only one dimension. Linear Equation - an equation of the first degree between two variables; -- so called because every such equation may be considered as representing a right line.

30. Nonlinear - Not in a straight line. Occurring as a result of an operation that is not linear. Containing a variable with an exponent other than one.

31. Function - A variable so related to another that for each value assumed by one there is a value determined for the other. A rule of correspondence between two sets such that there is a unique element in the second set assigned to each element in the first set.

32. Reasoning - The basis or motive for an action, decision, or conviction. An underlying fact or cause that provides logical sense for a premise or occurrence: *There is reason to believe that the accused did not commit this crime.* The capacity for logical, rational, and analytic thought; intelligence. A premise, usually the minor premise, of an argument

Strand 6 Terms

1. Probability - The quality or condition of being probable; likelihood. A number expressing the likelihood that a specific event will occur, expressed as the ratio of the number of actual occurrences to the number of possible occurrences.

2. Certainty - A fact or truth unquestionable established.

3. Event -Something that takes place; an occurrence. A significant occurrence or happening. The final result; the outcome.

4. Random - Having no specific pattern, purpose, or objective. Of or relating to a type of circumstance or event that is described by a probability distribution. Of or relating to an event in which all outcomes are equally likely

5. Variation - A partial change in the form, position, state, or qualities of a thing; modification; alternation; mutation; diversity; deviation; as, a variation of color in different lights; a variation in size; variation of language.

6. Event - Something that takes place; an occurrence. A significant occurrence or happening. The final result; the outcome.

7. Probabilistic - Of, based on, or affected by randomness, or chance.

8. Dependence - The state of being influenced and determined by something; subjection (as of an effect to its cause).

9. Independence - The state or quality of being independent; freedom from dependence; exemption from reliance on, or control by, others; self-subsistence or maintenance; direction of one's own affairs without interference.

10. Simulation - Attempting to predict aspects of the behavior of some system by creating an approximate (mathematical) model of it. This can be done by physical modeling, by writing a special-purpose computer program or using a more general simulation package, probably still aimed at a particular kind of simulation (e.g. structural engineering, fluid flow). Typical examples are aircraft flight simulators or electronic circuit simulators.

11. Empirical - Relying on or derived from observation or experiment. Verifiable or provable by means of observation or experiment. Guided by practical experience and not theory, especially in medicine.

12. Theoretical - Of, relating to, or based on theory. Restricted to theory; not practical. Given to theorizing; speculative.

13. Model - To plan or form after a pattern; to form in model; to form a model or pattern for; to shape; to mold; to fashion

14. Discrete - Defined for a finite or countable set of values; not continuous. Constituting a separate thing. Consisting of unconnected distinct parts

15. Scheduling - The arrangement of a number of related operations in time.

16. Sequence - A following of one thing after another; succession. An order of succession; an arrangement. A related or continuous series. An ordered set of quantities, as x, $2x^2$, $3x^3$, $4x^4$.

17. Network - An intricately connected system of things or people. An interconnected or intersecting configuration or system of components.

18. Recurrence - Happening again (especially at regular intervals).

19. Induction - A two-part method of proving a theorem involving an integral parameter. First the theorem is verified for the smallest admissible value of the integer. Then it is proven that if the theorem is true for any value of the integer, it is true for the next greater value.

20. Iteration - A computational procedure in which a cycle of operations is repeated, often to approximate the desired result more closely.

21. Algorithm - A step-by-step problem-solving procedure, especially an established, recursive computational procedure for solving a problem in a finite number of steps.

22. Permutation - A rearrangement of the elements of a set. A complete change; a transformation. The act of altering a given set of objects in a group.

23. Combination - One or more elements selected from a set without regard to the order of selection. The act of combining or the state of being combined. The result of combining.

24. Set - A group of things. A collection of distinct elements having specific common properties: *a set of positive integers.*

25. Vertex edge graph – A finite nonempty set of vertices, along with a set of edges.

26. Critical paths – The minimum time needed to complete a task.

27. Minimum spanning trees – A connected graph that has the following properties 1) no cycles, 2) exactly one path between any two vertices, 3) for more than one vertex, the tree will have at least two vertices of degree 1, 4) with n vertices, the tree will have n-1 edges and 5) one spanning tree must have a weight that is less than or equal to the weight of all the rest.

28. Adjacency matrix – For any graph G(a,b), the adjacency matrix is an nxn matrix and the entry in the ith row and j^{th} column of the matrix represents the number of edges from V_i to V_j.

29.Optimization problems – Problems involving finding the maximum or minimum of an entity.

Answer Key

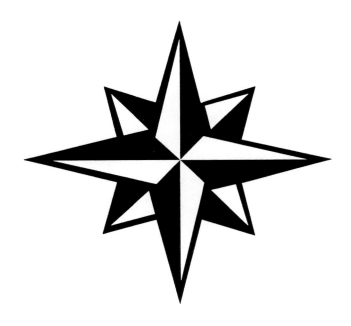

Problem Number	Strand 1 1.1
1	B
2	C
3	C
4	D

5.

Week X	Red Maples Y
1	150
2	230
3	310
4	390
5	470

I will use my TI-83 to find the relationship that best fits (linear regression). Since the difference in successive "Y" terms is constant at the first level, I knew the equation was linear.

$Y = 80x + 70$

This means that when I subtract each "Y" term from the next, if I see the same number or difference then the equation is linear. This solution strategy is called "using finite differences".

6.

Week X	Silver Maples Y
1	50
2	125
3	250
4	425
5	650

I will use my TI-83 to find the relationship that best fits (quadratic regression). Since the difference in successive "Y" terms is constant on the second level, I knew the equation was quadratic.

$Y = 25x^2 + 0x + 25$

Notice that this time I subtracted twice or had to go to the second level.
If you find your constant on the first level (or when subtracting once) then the equation is linear.
If you find your constant when subtracting twice (or on the second level) then the equation is quadratic.
If you find your constant when subtracting three times (or on the third level) then the equation is cubic...........

7. $25x^2 + 0x + 25 = 80x + 70$
$\quad 25x^2 - 80x - 45 = 0$

I used the Quadratic Formula and my TI-83.

A = 25, B = -80 and C = -45
X = 3.69 and -.488
Since you want to know the number of weeks in the future, the answer is 3.69 weeks or, between the third and fourth week.

8.

Class	Number of Roses Sold
Freshmen	55
Sophomores	83
Juniors	112
Seniors	72

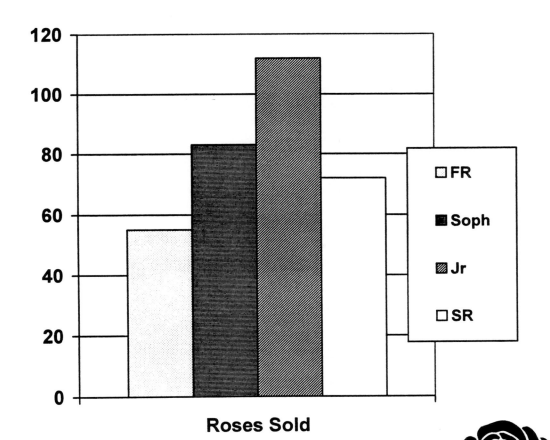

Roses Sold

Percent = Part/Total
Freshmen = 55/322 = 17%
Sophomores = 83/322 = 26%
Juniors = 112/322 = 35%
Seniors = 72/322 = 23%

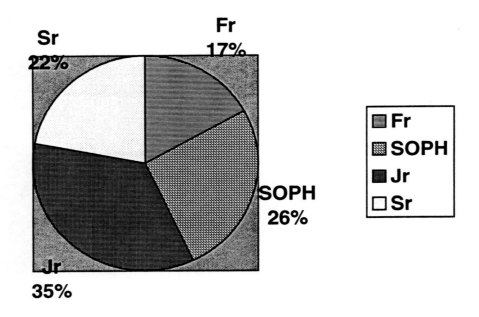

Problem Number	Strand 1 1.2
1	B
2	C
3	C
4	D

5.

Level	Length	Width	Perimeter	Area
1	1	1	4	1
2	2	2	8	4
3	3	3	12	9
4	4	4	16	16

Length – How long each figure is
Width – How wide each figure is
Perimeter – The distance around each figure
Area – The number of square units it took to make up each figure

-Or-

Level	Length	Width	Perimeter $P=2l+2w$	Area $A=s^2=l^2=w^2$
1	1	1	$2(1)+2(1)=2+2=4$	$l^2=w^2=1^2=1$
2	2	2	$2(2)+2(2)=4+4=8$	$l^2=w^2=2^2=4$
3	3	3	$2(3)+2(3)=6+6=12$	$l^2=w^2=3^2=9$
4	4	4	$2(4)+2(4)=8+8=16$	$l^2=w^2=4^2=16$

Length	Perimeter
1	4
2	8
3	12
4	16

6. Since the difference in successive "Y" terms is constant at the first level, the relationship is linear. Perimeter is one-dimensional concept hence it is linear. I used my TI-83 to find the line of best fit (linear regression).

$Y = 4x$ or $P = 4L$
For an object with a length of 10 units
$P = 4L = 4(10) = 40$ units.

7.

Length	Area
1	1
2	4
3	9
4	16

Since the difference in successive "Y" terms is constant at the second level, the relationship is quadratic. Area is two-dimensional concept hence it is quadratic. I used my TI-83 to perform a quadratic regression.

$Y = ax^2 + bx + c$
$Y = 1x^2 + 0x + 0$
$Y = 1x^2$ Or $A = L^2 = W^2$

For an object with a length of 10 units

$A = L^2 = W^2 = 10^2 = 100$ square units.

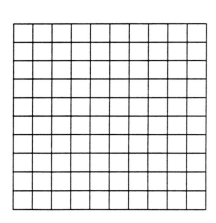

8. The equations for Perimeter and Area apply to squares.

P = 4L = 4(12) = 48 units (48 units around)

A = L² = (12)² = 144 square units (144square units in the total figure)

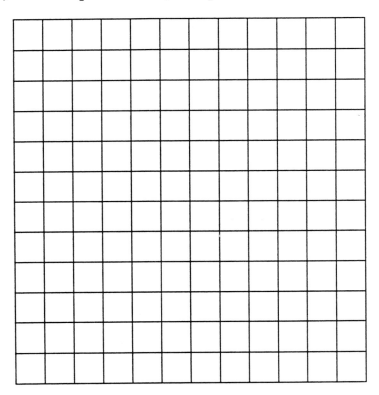

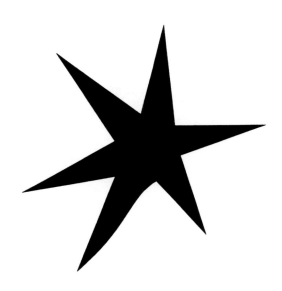

Problem Number	Strand 1 1.3
1	A
2	D
3	B
4	A
5	A
6	C

7. New Length = 40 + (40 x .50) = 40 + 20 = 60 feet
New Width = 30 (30 x 1.00) = 30 + 30 = 60 feet

Old Area
(Old Length) x (Old Width) =
40 x 30 =
1200 square feet

New Area
(New Length) x (New Width) =
60 x 60 =
3600 square feet

Percent Increase =
Amount of Increase/Original Amount =
2400/1200 = 2.00 or
A 200% increase in the area of the Kiddie Pool.

Old Perimeter

2(Old Length) + 2(Old Width) =
2(40) + 2(30) =
80 + 60 =
160 feet

New Perimeter

2(New Length) + 2(New Width) =
2(60) + 2(60)
120 + 120
240 feet

Percent Increase =

Amount of Increase/Original Amount =
100/140 =
.714 or
A 71.4% increase in the Perimeter of the Kiddie Pool.

8. Using the percent increases from number 7, the new pool has an area of 3600 square feet. The Kiddie Pool should generate

$7.50 x 3600 or $27,000 per day.

Problem Number	Strand 2 2.1
1	B
2	D
3	C
4	D
5	C
6	D
7	C

8. I used the "Y = " and "Table" features on my TI-83 to generate these values.

X	Y
1	155
2	210
3	265
4	320
5	375

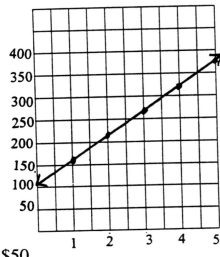

New Service Fee = 100 – (100x.50) = 100 – 50 = $50

Hence, Y = 55x + 50. All the corresponding "Y" values are reduced 50 units so the graph is lowered 50 units.

X	Y
1	105
2	160
3	215
4	270
5	325

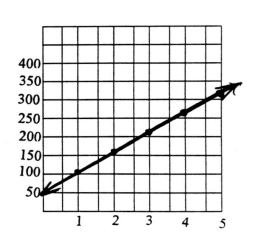

Problem Number	Strand 2 2.2
1	B
2	B
3	B
4	D
5	D
6	A
7	D

8. $V = \Pi r^2 h$ Surface Area = Circumference of Base x h + $2\Pi r^2$

I used my TI-83 with the formulas for the volume and surface area of a cylinder.

(r = 2 & h = 14)

V = 175.93 cm³   **SA = 201.06 cm²**

203

Problem Number	Strand 2 2.3
1	A
2	B
3	C
4	C
5	A
6	D
7	B

8. Since 3 feet = 1 inch

$$\frac{3ft}{X} = \frac{1in}{5in}$$

X = 15ft

$$\frac{3ft}{X} = \frac{1in}{8in}$$

X = 24ft

The actual size of the dining room is 15ft x 24ft or **360 square feet**.

Problem Number	Strand 3 3.1
1	D
2	B
3	C
4	C
5	C
6	A
7	C

8.
Supplies $20
Gasoline $12
Trash Bags $15
Total $47

$162 – 47 = $115
(Dave and Carl split $115 which means they both received $57.50)

Supplies = 20/162 = 12%
Gas = 12/162 = 7%
Trash Bags = 15/162 = 9%
Dave = 57.50/162 = 35%
Carl = 57.50/162 = 35%

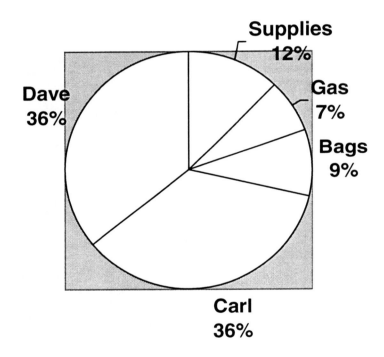

Problem Number	Strand 3 3.2
1	D
2	D
3	C
4	D
5	D
6	C
7	D

8. The median or the middle number would be more representative. Most students passed the test with a 70 or higher. The "0" and the "12" are skewing the class mean (Median=71 & Class Mean = 61).

Stem	Leaf
0	0
1	2
5	4
6	5
7	0,2,8
8	0,4
9	1

Problem Number	Strand 3 3.3
1	A
2	C
3	A
4	B
5	D
6	D
7	D

8. a)

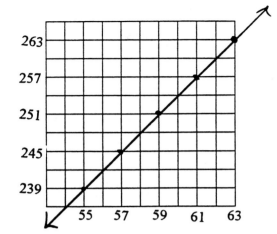

X	Y
55	239
57	245
59	251
61	257
63	263

b) Since the difference in successive "Y" terms is constant at the first level, the relationship is linear.

I used my TI-83 to perform a linear regression

$$Y = 3x + 74$$

c) $\dfrac{239 + 245 + 251 + 257 + 263}{5} = $ **251cm**

d) Y = 3x + 74

   3(48) + 74

   218cm

209

Problem Number	Strand 4 4.1
1	D
2	D
3	B
4	C
5	D
6	B
7	C

8.
Amy = 1/2 = 50%
Beth = 1/4 = 25%
Cam = 1/8 = 13%
Dee = 1/16 = 6%
Eve = 1/32 = 3%
Fred = 1/32 = 3%

Amy = .5 x 684.26 = $342.13
Beth = .25 x 684.26 = $171.07
Cam = .125 x 684.26 = $85.53
Dee = .0625 x 684.26 = $42.77
Eve = .03125 x 684.26 = $21.38
Fred = .03125 x 684.26 = $21.38

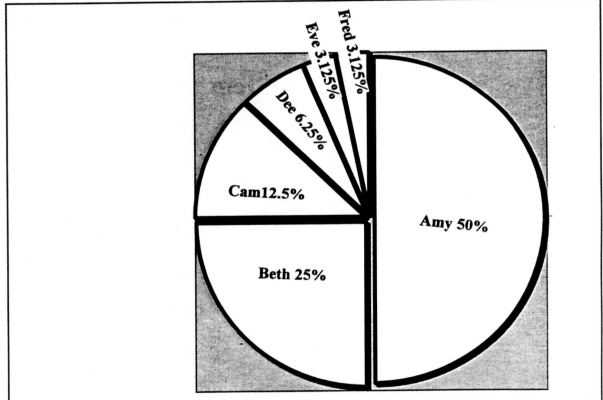

Problem Number	Strand 4 4.2
1	B
2	D
3	D
4	D
5	A
6	D
7	A

8. a) Distance = Rate x Time
Time = Distance/Rate
Time = 250/55
Time = 4.55 hours

b) Distance = Rate x Time
Rate = Distance/Time
Rate = 250/6
Rate = 41.67 miles per hour

Problem Number	Strand 4 4.3
1	A
2	B
3	C
4	C
5	C
6	B
7	C

8. a) 334 x .20 = $66.80 for Tithes and Offerings
When Shay deducts 20% for Tithes and Offerings she will have
$334.00 - $66.80 = **$267.20**

b) When Shay gives $50 to her sister she will have given her
50/267.20 = **18.71%**

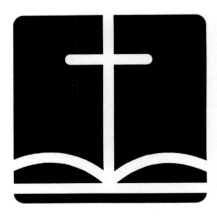

Problem Number	Strand 5 5.1
1	C
2	A
3	B
4	B
5	D
6	B
7	A

8. C – Number of Chocolate Donuts
G – Number of Glazed Donuts

$C + G = 295$
$\underline{.75C + .5G = 200}$
$C = 295 - G$
$\underline{.75C = .5G = 200}$
$.75(295 - G) + .5G = 200$
$221.25 - .75G + .5G = 200$
$221.25 - .25G = 200$
$-.25G = -21.25$
$G = 85$ (The Senior Class sold 85 Glazed Donuts)

Substitution

Since $C + G = 295$
$C + (85) = 295$
$C = 210$ (The Senior Class sold 210 Chocolate Donuts)

Problem Number	Strand 5 5.2
1	A
2	C
3	A
4	D
5	A
6	D
7	C

8. D – Number of Dimes
Q – Number of Quarters

$D + Q = 100$
$.10D + .25Q = 21.40$
$\quad\quad D = 100 - Q$
$.10D + .25Q = 21.40$
$.10(100 - Q) + .25Q = 21.40$
$10 - .10Q + .25Q = 21.40$
$10 + .15Q = 21.40$
$.15Q = 11.40$
$Q = 76$ (Alonzo has 76 quarters)

Since $D + Q = 100$
$\quad\quad D + (76) = 100$
$\quad\quad D = 24$ (Alonzo has 24 Dimes)

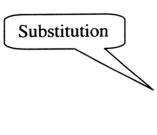

Substitution

Problem Number	Strand 5 5.3
1	D
2	C
3	D
4	D
5	A
6	D
7	B

8. L = Length
W = Width

$L=2W+4$

$\underline{2L+2W=56}$

$2(2w+4)+2W=56$ Substitution

$4W+8+2W=56$

$6W+8=56$

$6W=48$

W=8 (Kelly's room is 8cm wide)

Since L = 2W + 4

$L=2(8)+4$

$L=16+4$

L=20 (Kelly's room is 20cm long)

Problem Number	Strand 6 6.1
1	B
2	D
3	C
4	A
5	C
6	A
7	A

8.a) Matt/Lu/Ra
Matt/Lu/Le
Matt/Lu/Na
Matt/Jn/Ra
Matt/Jn/Le
Matt/Jn/Na
Matt/Ru/Ra
Matt/Ru/Le
<u>Matt/Ru/Na</u>
Mark/Lu/Ra
Mark/Lu/Le
Mark/Lu/Na
Mark/Jn/Ra
Mark/Jn/Le
Mark/Jn/Na
Mark/Ru/Ra
Mark/Ru/Le
<u>Mark/Ru/Na</u>

Dav/Lu/Ra
Dav/Lu/Le
Dav/Lu/Na
Dav/Jn/Ra
Dav/Jn/Le
Dav/Jn/Na
Dav/Ru/Ra
Dav/Ru/Le
<u>Dav/Ru/Na</u>
Esth/Lu/Ra
Esth/Lu/Le
Esth/Lu/Na
Esth/Jn/Ra
Esth/Jn/Le
Esth/Jn/Na
Esth/Ru/Ra
Esth/Ru/Le
Esth/Ru/Na

b) Esth/Ru/Ra
Esth/Ru/Le
Esth/Ru/Na

P=3/36=1/12

Problem Number	Strand 6 6.2
1	D
2	C
3	C
4	A
5	D
6	D
7	C

8. a)Members of both sets, {Ezekiel, Paul, David, Karen, Saul and Tom}

b) Members of either set {Victor, Ezekiel, Paul, David, Karen, Saul, Tom, Tim, Ralph, Walter, Harold, and Sharome}

Problem Number	Strand 6 6.3
1	C
2	C
3	C
4	B
5	D
6	D
7	D

8.a) Alice/David,
Alice/Shadrach,
Alice/Meschach
Alice/Abednego
Beth/David,
Beth/Shadrach,
Beth/Meschach
Beth/Abednego
Carmen/David,
Carmen/Shadrach,
Carmen/Meschach
Carmen/Abednego
Dana/David,
Dana/Shadrach,
Dana/Meschach
Dana/Abednego

b) Alice/Abednego
   Dana/David

P=2/16=1/8

The MEAP Solution:

A Technological Approach

Written by
Twianie Mathis, Ed.D

ISBN 0-9718019-0-8: $25.00

Mathematical Solutions Publishing Company
P.O. Box 36365
Grosse Pointe Farms, Michigan 48236-0365

Mathematical Solutions Publishing Company

would like to hear from you. If you have any comments or suggestions, please indicate them below and mail this form to:

> Mathematical Solutions Publishing Company
> PO Box 36365
> Grosse Pointe Farms, Michigan
> 48236-0365

Comments/ Suggestions

_____

_____

_____

_____

_____

_____

The MEAP Solution:
A Technological Approach

TI-83 Edition
9-12 Math

Order Form

- 9-12 Math
- Promotes technology
- Multiple Choice and Constructed Response Questions
- Material Presented By Objective
- TI-83 programs correspond to MEAP reference sheet
- Easy step- by- step assistance provided
- Easy To Use Class Worksheets
- Use as part of your S.I.P.

Please Send Purchase Order, Check or Money Order to
Mathematical Solutions Publishing Company
P.O. Box 36365
Grosse Pointe Farms, Michigan 48236-0365
Or
For Faster Service
Fax Your Purchase Order To (313) 881-2477

Item	Cost
"The MEAP Solution: A Technological Approach" $750.00 **sold in class sets of 30** ($25.00 per book)	_____ x $750.00 = _____
Tax (6.00%)	
Shipping and Handling (10%)	
Total	